# When Words Betray Us

Sheila E. Blumstein

# When Words Betray Us

Language, the Brain, and Aphasia

 Springer

Sheila E. Blumstein
Cognitive, Linguistic & Psychological Sciences
Brown University
Providence, RI, USA

ISBN 978-3-030-95850-3          ISBN 978-3-030-95848-0    (eBook)
https://doi.org/10.1007/978-3-030-95848-0

This Springer imprint is published by the registered company Springer Nature Switzerland AG
The registered company address is: Gewerbestrasse 11, 6330 Cham, Switzerland

*To all those, here and gone, who made a difference in my life*

**The ASA Press**

ASA Press, which represents a collaboration between the Acoustical Society of America and Springer Nature, is dedicated to encouraging the publication of important new books as well as the distribution of classic titles in acoustics. These titles, published under a dual ASA Press/Springer imprint, are intended to reflect the full range of research in acoustics. ASA Press titles can include all types of books that Springer publishes, and may appear in any appropriate Springer book series.

*Editorial Board*

# Prologue

My first introduction to aphasia was personal. At the time I did not know what aphasia was. When I was 19, my grandmother had a stroke. Within days of her stroke, I visited her in her home where she was lying in a hospital bed in her room. She could no longer speak. Although she tried, what came out were incomprehensible syllables. Over time, I could tell from my visits that she was improving. But the improvement was painfully slow. She could not walk on her own, she could say only a few words, and it was not clear how much she understood. Unfortunately, her story did not have a happy ending. She continued to have small strokes, and as a result, her language declined until she ultimately stopped all communication.

At the time, no one gave a name to her disorder. It was not until I was in graduate school that I made the connection when one of my professors, Roman Jakobson, talked about the breakdown of language in adults as a result of brain injury. I realized that my grandmother had had aphasia. It was then that I started reading and wrote my first paper in that course about aphasia. My grandmother may have lost her voice, but it has lived on as her legacy to me in my lifelong interest and commitment to understanding language, the brain, and aphasia.

Having been in the field since 1967 (starting in my second year as a graduate student), one has what I call the long view. Theories come and go, old theories are rediscovered, and new theories are developed. Progress seems infinitely slow – but with the long view, one can see that there is real and substantial change. And one also develops a personal perspective. And that is what this book represents. It is not intended to be a historical overview nor a critical analysis of the research of the field, although my worldview has been influenced by and owes much to those before me. It was written with the goal of making it accessible to the non-professional who may have personal experience with someone with aphasia or who is simply interested in learning about how language works, how it breaks down, and how the machinery of the brain makes it all happen.

There are many to thank for the support I have had throughout the years and who have helped bring this book to fruition. The list of people included here is not exhaustive, but it represents those who have deeply influenced and assisted me. First to my mentors of years ago, Roman Jakobson and Harold Goodglass. They provided guidance and support as I learned about aphasia and started my research career. In preparing this book, special thanks go to Carol Fowler and Donald Shankweiler who read drafts of all of the chapters and provided many helpful comments and criticisms; to Ann Marie Clarkson, a talented artist, who drew many of the figures included in the book; and to my colleagues David Badre, Elena Festa, Bill Heindel, and Philip Lieberman and former students, now colleagues, Sara Guediche, Allard Jongman, Sahil Luthra, Emily Myers, Rachel Theodore, and Kathleen Kurowski for many helpful discussions and assistance. Thanks to the *Labites*, the undergraduate and graduate students and the postdocs, who were part of my lab at Brown University. Many thanks go to the team at Springer and especially to Sam Harrison, a great editor, who gave consistently helpful comments and advice. Finally, my gratitude goes to the many persons with aphasia with whom I worked and who were gracious participants in my research program and also to the National Institutes of Health and the National Institute on Deafness and Other Communication Disorders for supporting my research starting with a predoctoral fellowship in 1968 and continuing throughout my career.

# Contents

# Introduction

**1**

Think of the times when you are talking to someone and a word that you want to say escapes you, the perfect word that expresses your thought. Sometimes you never come up with the word, or other times the word just pops into your head, often hours later, unrelated to what you are doing, with neither conscious thought nor prompting. Even when you fail, you are sure you 'know' the word that you are searching for. You may be able to identify the first letter of the word, perhaps know how many syllables it has, or be able to name a word that is similar to it. Indeed, the word feels like it is right there *'on the tip of your tongue'*.

Or imagine your embarrassment, when you are at a party and a friend asks who that woman is standing next to your father, and instead of saying it is your aunt, you say it is your mother. Oops – how did that uninvited word, that *'slip of the tongue'* come out?

Words betray us all of the time – they can be frustratingly elusive; yet fortunately, for most of us, these misnomers occur relatively infrequently. Indeed, we just assume the normalcy of language, taking it for granted. It is always there for us, except for those few exceptional, interesting, and sometimes humorous lapses. But for some, such lapses are a constant feature of life. And difficulty in finding the right word may be only one problem or *symptom* that these individuals or persons must endure every day and often every moment. They may have a host of other problems in using language. Not just selecting words that form a sentence, but articulating them, stringing words together in a sentence, understanding what others are saying, and using the literary arts including reading and writing. Such individuals have aphasia.

How does this happen? Persons with aphasia have typically had a stroke or some brain injury as adults that have compromised their speaking and understanding of language. Prior to their neural episode, they were just like you and me – using language daily and never thinking about it. But that can change in an instant, and the consequence is that, on a daily basis, a person with aphasia is now fully cognizant that using language – saying whatever comes to mind, expressing feelings, ideas, understanding what others are saying – is compromised at best, impossible at worst.

© Springer Nature Switzerland AG 2022
S. E. Blumstein, *When Words Betray Us*,
https://doi.org/10.1007/978-3-030-95848-0_1

Think about the devastation that would cause. Language is the window into who and what we are. Language perhaps more than any other cognitive function defines us as human beings. It provides our lives with richness and defines not only our culture but also who we are as individuals. It serves as the primary vehicle for interacting with others. Language is the connection between our inner self and the world around us, and it provides the vehicle for us to be social and productive human beings and to navigate the world.

The goal of this book is to tell the story of aphasia – the what, the how, and the why. What is the nature of the deficits that aphasics have, how does the brain put together the pieces of language into a unitary whole, and why do particular areas of the brain underlie language and its many components? As with all good stories, there are multiple threads that comprise it. There is the story itself; besides witnessing the human toll of this disorder, the story of aphasia gives us a window into language and brain function. It allows for a rich picture that elucidates a tapestry of spared and impaired language abilities, and the complexity of the mapping of these abilities on to our neural machinery.

Elucidating the relation between language and the brain has its origins in neuropsychology – a field whose goal is to understand the neural bases of cognitive functions by studying the behavior of individuals who display different impairments pursuant to brain injury. Neuropsychology has a long and rich history, and the study of persons with aphasia has been a major part of it. In the early years, from the late nineteenth century to the mid-twentieth century, there were rich descriptions of persons with aphasia made by neurologists, psychologists, and neuropsychologists such as Henry Head, Kurt Goldstein, Hughlings Jackson, and Alexander Luria, to name a few. It was shown that although the two (left and right) hemispheres of the brain are structurally symmetrical, they are not functionally symmetrical. Behavioral tests of patients with injury to either their left or right hemisphere indicated that the two hemispheres do different things. The critical role of the left hemisphere in language was established – left hemisphere brain injury resulted in aphasia for most adults, whereas damage to the right hemisphere did not.

The complexity of the organization of language in the left hemisphere was also established. Based on observations and descriptions of individuals who sustained brain injury and on analyses of post-mortem autopsies, it was shown that damage to different areas of the left hemisphere resulted in different patterns of impairment – with a complex interaction of both largely spared and severely impaired language abilities. For example, as we will describe in detail in Chap. 2, Broca's aphasics with damage affecting frontal structures of the left hemisphere understand language well and yet show impairments in expressing language, whereas Wernicke's aphasics with damage to temporal lobe structures of the left hemisphere have difficulty understanding language and yet produce language easily and fluently. These constellations of spared and impaired abilities or *symptom-complexes* provided the basis for hypotheses about the function of lesioned areas of the brain.

Hypotheses based on descriptions are just the start of scientific investigations. Testing of these hypotheses is the next step and requires application of rigorous experimental methods. The history of aphasia is no different. Starting in the mid-1960s, this psycholinguistic or neurolinguistic approach was the dominant means of studying language and the brain. Here, a range of experimental paradigms, largely drawn from psychology, were used to increase the understanding of aphasics' language behavior in order to try to explain the basis of the deficit coupled with an explanation of the function of the damaged neural tissue. For example, consider what your mind/brain has to do to understand a word such as 'cat'. You have to perceive the sounds of the word, 'c', 'a', 't'. The sounds have to match a particular word from among all of the words in English that you know (it's 'cat' not 'bat', 'cut' or 'cap'). You have to select that word, 'cat', and map its sounds to its meaning. So if an aphasic does not understand a word, the question is why? Does a failure to understand the meaning of 'cat' reflect a problem in processing sounds, matching sounds to words, mapping words to meanings, or unpacking the meanings of words themselves? A series of experiments would need to be designed and run, and then related to where the brain injury is. Do persons with damage in this area typically show problems with understanding words? And if so, do they display similar results across experiments that use different methods? Indeed, the bedrock of our knowledge of the neural bases of language comes from these experimental studies of aphasia. In the chapters that follow, we will explore not just the effects of brain injury on language behavior through the lens of aphasia but also examine what these findings tell us about how language is processed in the brain as we speak, understand, and communicate with others.

The importance of this endeavor is multifold. It provides a window into the history of an exciting and ever evolving science. Indeed, it reflects the incrementalism of science; while there are changes and often seemingly entirely new insights and discoveries, most have deep historical roots. We will see how the basic elements remain, but we will also see how technological advances provided by brain imaging and computer-based modeling have helped shape our current knowledge of language and the brain. These methods allow for a precise mapping of the injured brain that was difficult, if not impossible, in earlier times. Their application to persons with brain injury as well as to the uninjured brain allows us to compare language in the brain when it is damaged to when it is spared from injury. Do lesions in aphasia predict the areas that give rise to language in the uninjured brain?

We will also see how computational modeling has shaped how we look at language and its neural basis. Is the brain like a computer? And what type of computer? We will see how current computational models have properties that mirror those of neurons, thus providing a biologically-driven theoretical framework as we examine how the brain processes information about the components of language and integrates these pieces into a unitary whole as we speak and understand. Is language broken up in the brain into neural regions that are specialized for particular linguistic functions or modules such as speech, words, syntax, and semantics, or is language represented in a broadly distributed network of connections (similar to networks of neurons) where each linguistic function recruits multiple neural areas that have different but complementary functions?

And what happens to our computational model when a part of it is damaged? What happens to the real brain under these circumstances? What happens to language in aphasia? Can the adult brain recover from or adapt to injury? In other words, does the brain show *plasticity*? We can examine this question by looking at whether and how persons with aphasia recover language, and what brain regions support such recovery.

We can also examine brain plasticity by looking at what happens to areas of the brain that are involved in language when these areas are deprived of input. We use our ears to process the sounds of language. What happens to those neural areas if deprived of sound input, as is the case for people who are deaf? The deaf use an alternate language (in the United States, they use American Sign Language), allowing them to communicate by eye rather than by ear (for understanding) and by hand rather than by mouth (for speaking). Do the deaf use different neural areas for language than those who are hearing?

The same question can be asked for the blind who can no longer read by eye, but rather read by touch using Braille. What happens to those neural areas that the sighted use for reading if they are deprived of input because of blindness?

Much discussion thus far has focused on the critical role that the left hemisphere plays in language processing. But there is a right hemisphere which has analogous structures to the left. What role, if any, does it play in language processing? Is it waiting in the wings to take over the functions of the left hemisphere when damaged? Does it have the capacity to pinch-hit for the damaged left hemisphere? Or does it have its own set of functions that contribute to the unity of language?

At its center, this book is about the science of aphasia. But it is also about the human toll of this language disorder on the life of those with aphasia and the impact this has on their families. Every case is unique, but the constant is that life has changed inexorably; nothing is the same as before. Is there a road map to recovery from aphasia? What are the challenges? Answers to these questions are also a part of our story.

So let us start our journey and see how language meets brain. We will begin with some basics on aphasia and properties of language and the brain. Taken together, they provide the framework that guides the remainder of this book as we seek to understand what happens **When Words Betray Us**.

# Getting Started

**2**

## 2.1    The Study of Aphasia: The Breakdown of Language

*A1. 'I know what I want to say but I can't say it'*
*A2. 'boy...two...crying...sad sad...'*
*A3. P$^{hhh}$uppy... car hit... the leg is blood... fet... no vet...the puppy...OK*
*A4. 'I saw it and it was so that I knew it to be that it was'*
*A5. 'Every tay (day) I saw spalika when it had to'*

These are some examples of what you might hear if you were talking to different people with aphasia. As you can see, there is a variety of problems, each somewhat different. A1 is well aware of the problem he has in accessing the words he wants to use. A2 is quite different. A2's speech comes out in a halting fashion, with pauses (…) interspersed between most words. Also, the 'little' grammatical words of English, such as 'the', 'is', are lacking. Yet despite this, meaning is conveyed fairly well, describing that there are two boys, and either one or both are *very* sad. The degree of sadness is accomplished without using the adverb *very* but instead by repeating the word 'sad' to intensify the degree of the mood.

A3 also tells a short but comprehensible story. Pauses are made between phrases. Speech sound errors occur: the production of the first sound in the word 'puppy' is pronounced with a strong burst of breath as it is released ('p$^{hhh}$'), and 'v' in the word 'vet' is incorrectly produced as 'f'. In the latter example, A3 corrects the production, indicating an awareness that an error had been made. In contrast to A2, A3 uses the grammatical words, 'the' and 'is', although 'is' is used incorrectly.

A4 and A5 are totally different. Both are fluent in their speech output. They produce a lot of words with no pauses interspersed. However, it is not clear what each of them is trying to say. A5 produces the word 'day' incorrectly as 'tay'. What about 'spalika'? Could it be a word you never heard of in English? Nope. It sounds like an English word, but it is not.

Considering these examples, it looks like the breakdown of language can happen willy-nilly. This is not true, however. There is a pattern to the various language deficits shown above. This pattern reflects the intersection of two important principles

S. E. Blumstein, *When Words Betray Us*,
https://doi.org/10.1007/978-3-030-95848-0_2

that we will be following throughout this book. One is the *architecture of language* (how the pieces of language are put together) and the other is the *functional architecture of the brain* (how different areas of the brain support language and its various parts).

In this chapter, we will introduce aphasia by describing the different clinical syndromes (behavioral manifestations in using language) of this disorder. We will tie this in with some basic neuroanatomy and consider whether there is functional specialization of areas of the brain for language and its parts. We will then look at the components of language and examine how they are put together. To do so, we introduce a computational framework (called *connectionist*) which attempts to explain how linguistic information is represented and passed through the system in a biologically/neurologically plausible way. Here, we will see how language meets the brain as we begin our study of aphasia.

## 2.2    Some Preliminaries

Aphasia is not rare. It usually occurs as the result of a stroke, but there are other ways that people can get aphasia including incurring a head trauma, having a brain tumor, or a brain bleed. According to the National Aphasia Association (https://www.aphasia.org/), in the United States there are over 2,000,000 Americans who have aphasia with about 180,000 acquiring aphasia annually. These are sobering numbers indeed. That said, it is important to emphasize that aphasia is neither a regression to an earlier stage of language nor is it a cognitive impairment. This should dispel common misconceptions that someone who has a language impairment is no longer able to reason, think, have opinions, feelings, or interact normally with the world around them. Not true. Those who have aphasia are language impaired; they have not reverted to an earlier stage of language nor are they intellectually compromised.

As the introductory examples at the beginning of this chapter suggest, there is not one aphasia. Rather, there are multiple types of aphasia or *syndromes* (a group of symptoms that occur together) which occur as a result of brain injury to different parts of the left hemisphere. Meeting someone with aphasia and striking up a conversation gives a first-order picture of where there are difficulties in communicating and using language. Critical in such an assessment is looking at the 'two sides of the coin' that are critical to language communication – *language output* – can the person produce the sounds and words of language and does what is said make sense, and *language input* – can the person understand language and participate in a conversation. Clinical assessments of aphasia do just that, systematically looking at a range of communication behaviors and assessing both where things are wrong (disabilities) and also where thing are right (abilities).

It may seem strange to include abilities as well as disabilities in assessing a person with aphasia. The study of aphasia is not just about impairments. It is obviously critical to know what aspects of language are pathological. However, it is also essential to understand what aspects of language are spared or minimally affected.

As we shall see, there is much of language that is spared, even for those who have limited language production and/or for those who have impaired language comprehension. Thus, aphasia provides a nuanced picture of how language breaks down, presenting a rich tapestry of language behavior as a result of brain injury. Such information not only tells us about the language system and how it functions in our neural machinery, but it also provides potential practical applications for developing language therapy programs. Admittedly, this book is not a handbook on rehabilitation. But the findings we will talk about suggest a potential strategy for going from *bench* (the lab) to *bedside* (the person with aphasia): using what is 'right' (spared) may serve as a scaffold to 'bootstrap' and help (re)-build those aspects of language that are 'wrong' (impaired) (for discussion see Chap. 9).

## 2.3   Aphasia Syndromes

Think about what you would expect when you meet a friend. You would likely start with conversational 'niceties' like a greeting, e.g. 'Hi. Great to see you'; a question about how the person is, e.g. 'how are you'; and perhaps remarks on the weather 'it is freezing out there'. Then the conversation narrows to a particular topic such as the news, politics, a recent social gathering, each requiring the selection of words appropriate to that topic. Your expectations are that in producing language your friend will speak fluently, speak clearly, not make many, if any, slips of the tongue, make sense, stay on topic, and interact easily with you, responding appropriately to what you are saying. You also expect that your friend will understand what you are saying, and respond to the topic accordingly. If you do not pick up on a particular word, you might ask what was just said, and your friend will likely repeat that word or do a paraphrase of it. These aspects of language are among those looked at systematically in a clinical assessment of someone who has aphasia.

We now turn to the description of two aphasia syndromes, Broca's and Wernicke's aphasia. These two types of aphasia have been the subject of the most intensive study aimed at understanding how the components of language break down and what neural systems underlie the deficits.

### 2.3.1   Broca's Aphasia

There are characteristic features of those who are diagnosed with Broca's aphasia. Typical examples of their speech output are shown in the short speech samples of A2 and A3 presented in the beginning of the chapter. Speech output tends to be slow, labored, and difficult, typically with pauses between words. There are often sound errors, many of which are distorted. In addition, there are difficulties with syntax exemplified by either dropping grammatical words (A2) or using them, sometimes incorrectly (A3). Grammatical endings on words that serve to mark verb tense or object number may also be dropped. A2 does not use the plural on 'boy' to indicate that there are two boys. However, grammatical endings are not always omitted; note that A2 uses the grammatical particle '-ing' on 'cry'. These character-istic grammatical difficulties in producing language are called *agrammatism* or *tele-graphic speech* and are one of the hallmarks of Broca's aphasia. Nonetheless, despite the sparse nature of the output and the presence of agrammatism, speech production typically has semantic content and is understood by the listener.

Auditory comprehension tends to be good but not perfect in Broca's aphasia. Given objects or pictures to name, performance is generally good. Deficits begin to emerge when sentences become more grammatically complex and their meaning cannot be gleaned from real world probabilities. For example, a sentence like 'the hamburger is eaten by the boy' is understood because hamburgers don't eat boys. Similarly, 'the mouse is chased by the boy' is understood because mice don't typi-cally chase boys. However, a sentence like 'the midget is chased by the clown' is often misunderstood because real world probabilities do not dictate whether a midget chases a clown or a clown chases a midget. And sentences which have phrases embedded within them such as 'the woman who the man saw was carrying the suitcase' are problematic.

In a nutshell, those with Broca's aphasia have a speech output deficit affecting production of sounds and the use of syntax in the context of relatively good auditory comprehension. Despite their difficulty in carrying on a conversation because of speech output problems, they are aware and engaged in what is being said.

### 2.3.2  Wernicke's Aphasia

The short speech samples from A4 and A5 are emblematic of the speech output of Wernicke's aphasia. Compared to the samples of Broca's aphasia in A2 and A3, there is a lot more speech, and grammatical words and grammatical endings are present. Nonetheless, it is unclear what meaning is conveyed in A4 and A5 and such speech output is described as semantically empty. Those with Wernicke's aphasia also produce sound errors as shown by A5 who says 'tay' for 'day'. 'Spalika' is another 'word' that A5 produces. However, although it follows the sound pattern of English and in this sense is a *possible* word in English, there is no word 'spalika' in our vocabulary. Some, but not all, of those who have Wernicke's aphasia produce such *neologisms* or *jargon* (we will discuss the proposed origins of these neolo-gisms in Chap. 4).

These characteristic features of Wernicke's aphasia seem to reflect the opposite of what we saw with Broca's aphasia. In contrast to Broca's aphasia, speech output in Wernicke's aphasia is fluent, well-articulated, contains grammatical words and endings but is typically empty of semantic content. And in contrast to the relatively good auditory comprehension typical of Broca's aphasia, auditory comprehension is severely compromised. Comprehension deficits occur not only at the sentence level but even at the single word level.

Often, those with Wernicke's aphasia have what is called logorrhea or a press for speech. Logorrhea has its origins from Greek; *logo* meaning 'word' or 'speech' and *rhea* meaning 'flow' (as in 'diarrhea'). Clinically, Wernicke's aphasics with logorrhea talk incessantly and essentially 'run off at the mouth'. They just go on and on, appearing as if they do not get 'closure', a sense that they have made their point and it is now time for someone else to have a turn. Think of a conversation you might have with a friend. You say what you need or want to, and then let others reply or join in on the conversation. Often, the only way the examiner can stop someone who has logorrhea from continuing to speak is to hold up a hand as a cue that it is time to let others have a turn.

In sum, Wernicke's aphasia presents with fluent, well-articulated, but semantically empty speech coupled with poor auditory language comprehension. Unfortunately, poor auditory comprehension coupled with logorrhea and neologisms in speech output render meaningful communication extraordinarily difficult.

### 2.3.3    Summary of Clinical Syndromes

The different clinical pictures of Broca's and Wernicke's aphasia show that language fractionates in parts and is not represented in the left cerebral cortex as an indissoluble whole. Were the latter to be the case, we would see one aphasia syndrome with increased impairment emerging across the language spectrum as a function of the severity of the aphasia. Instead, the two syndromes are qualitatively different as shown in Table 2.1 by a dissociation between what is 'spared' and what is 'impaired' in language output and language comprehension.

Clinical assessment provides a first picture. However, there is much more to the story of aphasia. Clinical descriptions do not and are not meant to provide an explanation for the basis of the impairments nor how the brain processes and puts together the components of language. Knowing the areas of brain injury giving rise to Broca's and Wernicke's aphasia is essential. Indeed, looking only at the clinical syndromes and language behavior in general, absent any consideration of the brain, while interesting, fails to tell us about what areas of the brain give rise to the different syndromes and ultimately how language is mapped on to the brain. To begin the discussion, we provide a first introduction to some neural structures that will be critical for our discussion of language and the brain.

**Table 2.1**  Comparison of relatively spared (+) and impaired (−) language abilities in Broca's and Wernicke's aphasia

|  | Broca's Aphasia | Wernicke's Aphasia |
|---|---|---|
| **Language output** | | |
| Fluent | − | + |
| Articulation | − | + |
| Syntax | − | + |
| Meaningful | + | − |
| **Language comprehension** | | |
| Words | + | − |
| Sentences | + | − |

## 2.4    A Brief Introduction to the Brain

Most aphasias occur with left brain injury. As we described in Chap. 1, the brain is comprised of a right and a left hemisphere. The two hemispheres are anatomically symmetrical, having similar structures on the right and on the left. For most of us who are right-handers, language is represented in the left hemisphere. We know this because left brain injury in adults results in aphasia, whereas right brain injury typically does not. For this reason, we will focus our attention, at this point, on the left hemisphere, and consider the right hemisphere in Chap. 6.

Figure 2.1 shows a lateral view of the left hemisphere and identifies the lobes of the brain. The red circle circumscribes the areas within the major lobes of the brain that, when damaged, result in different patterns of language impairment. Lesions in the frontal lobe are often referred to as anterior lesions, and lesions in the temporal and parietal lobes are often referred to as posterior lesions.

You may wonder why the damage giving rise to aphasia seems to be circumscribed within the circled area. The answer is straightforward. It reflects how the blood supply feeds the cortex. The middle cerebral artery which courses along the sylvian fissure (see Fig. 2.1) is the source of oxygen and nutrients to the surface of the frontal, parietal, and temporal lobes. Just like a river, the main branch of the middle cerebral artery has branches or tributaries which feed areas both above the sylvian fissure (supra-sylvian) and below the sylvian fissure (sub-sylvian). A blood clot due to a stroke, for example, can occur at any point along the main branch or in any of the tributaries. It will impede or stop blood flow to areas beyond the clot leading to a reduction or a complete loss of oxygen and other nutrients to that area. The result will be injury or death to the neurons affected. Thus, most aphasias which are due to stroke occur in peri-sylvian areas, areas around the sylvian fissure.

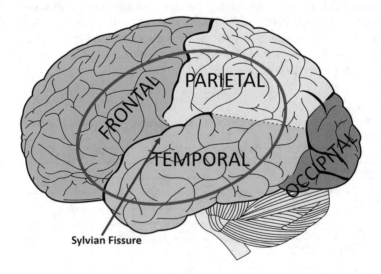

**Fig. 2.1**  Lateral (side) view of the left hemisphere

Lesions within these circumscribed areas give rise to a variety of aphasia syndromes in addition to Broca's and Wernicke's aphasia. These syndromes include conduction aphasia, anomic aphasia, transcortical motor aphasia, transcortical sensory aphasia, global aphasia, and alexia with agraphia. Each of these syndromes displays patterns of language breakdown that distinguish them from Broca's and Wernicke's aphasia and from each other. They provide additional evidence that there are functional differences in neural regions of the left hemisphere.

Indeed, the story of aphasia that we will tell in this book as well as evidence from more recent neuroimaging studies of non-brain-injured individuals discussed in Chap. 8 show that language and its components (sounds, words, sentences, meanings) are serviced by neural systems in the left hemisphere. What this means is that there is not one narrowly defined neural site involved in a particular language function. Rather, the processing of sounds, words, sentences, or meanings recruits multiple brain areas that are each a part of a broadly distributed neural system. We will investigate these neural systems in Chaps. 3, 4, 5 and 8.

Neural systems reflect how the various components of language are processed and ultimately recruited to accomplish a particular task. For example, as Fig. 2.2 shows, different neural areas are recruited when we hear words, see words, speak them, or are asked to come up with a word (in this case a verb) given a particular noun. Looking at the locus of neural activation, there are no surprises. The visual system and associated neural structures in the occipital lobe are activated when passively viewing words; the auditory system and associated neural structures in the temporal lobe are activated when passively listening to words; and the motor system

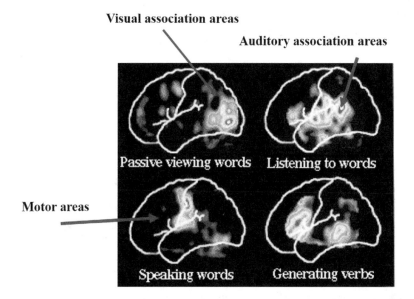

**Fig. 2.2** Neural areas activated during different language tasks. More active brain areas are represented by red and less activation by yellow and then green. Adapted from Posner, M.I. and Raichle, M.E. (1994). *Images of Mind.* Scientific American Library/Scientific American Books

in the frontal lobe is activated when speaking words. Of consequence, if you are given a noun such as 'dog' either auditorily or visually, and are asked to come up with a verb, it is necessary for you to access the word 'dog', its meaning, and then select an appropriate verb that relates to 'dog', such as 'bark'. As Fig. 2.2 shows, there is more extensive activation in this case than when passively reading, hearing, or saying a word. Importantly, you can see a broad range of neural structures that are recruited including a different area in the temporal lobe than the one activated when passively listening to words, and a different area of the frontal lobe than the one activated for simply repeating a word.

This example shows that the processing of words recruits multiple areas in a neurally distributed system located in the left hemisphere. Words are not processed in one narrowly localized 'place'. Nonetheless, as shown in this example, there are also neural areas that do have specialized functions related to seeing, hearing, and speaking – the functions that allow us to interact with the world.

### 2.4.1 Lesion Localization in Broca's and Wernicke's Aphasia

The different clinical pictures of Broca's and Wernicke's aphasia that we described earlier in this chapter suggest that there will be differences in the lesions resulting in their patterns of language breakdown. And indeed that is the case (Damasio 1998).

For Broca's aphasia, the lesion profile involves frontal areas of the left hemisphere (Damasio 1998). Damage is typically found in the inferior frontal gyrus (IFG) and may extend to adjacent premotor and motor areas (see Fig. 2.3). Additionally, there may be lesions to other areas not shown in Fig. 2.2 including the insular cortex (a part of the cerebral cortex located beneath the surface of the cortex)

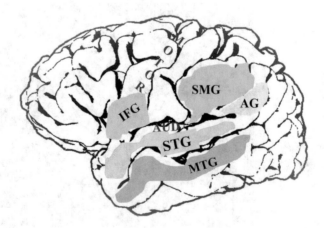

**Fig. 2.3** Neural areas in the left cerebral cortex involved in language processing. Areas include the inferior frontal gyrus (IFG), superior temporal gyrus (STG), middle temporal gyrus (MTG), supramarginal gyrus (SMG), and angular gyrus (AG). Speech input involves auditory areas (AUD) and speech output involves motor areas (MOTOR)

and the basal ganglia (white matter areas below the thin 1.5–5 mm thick layer of the cortex).

The lesion profile giving rise to Wernicke's aphasia involves temporal areas of the brain (Damasio 1998). It includes the posterior portion of the superior temporal gyrus (STG) but may extend along the length of the STG and also involves the middle temporal gyrus (MTG). It is not uncommon for the lesion to extend to the parietal lobe involving the supramarginal gyrus (SMG) and the angular gyrus (AG) (see Fig. 2.3).

### 2.4.2   Some Caveats on the Neural Localization of Syndromes

Consider that, as we described, there is a variety of neural areas that can be affected with a stroke in the middle cerebral artery. Add to that other causes of aphasia including trauma, tumors or bleeds. Given this scenario, it is highly unlikely that any two persons with aphasia will have exactly the same lesion, and they typically do not (Damasio 1998). In this sense, aphasia is the result of an 'experiment in nature' where the exact locus of the lesion and its extent can neither be controlled nor planned. This is the first challenge we have to face in looking at the effects of lesions on language behavior. In particular, there is variability in the sites of lesions giving rise to Broca's aphasia, on the one hand, and Wernicke's aphasia, on the other. To be sure, this variability does have limits in that, as we indicated above, those with Broca's aphasia have frontal lesions and those with Wernicke's aphasia have temporal or temporal/parietal lesions. Nonetheless, because the locus of the lesions may vary among patients who have the same syndrome, it is difficult to determine whether there is a one-to-one mapping between a particular, narrowly defined neural area and aphasia syndrome.

There are other challenges. Lesions producing aphasia tend to be large and thus extend to multiple neural areas. Indeed, small focal lesions typically produce a transient aphasia, one that disappears within a short period of time. This is a blessing for those individuals who have such an injury. For those who have a chronic aphasia, one that is long-lasting, lesions tend to be large and deep extending below the cortex to white matter subcortical structures. As multiple areas may be lesioned, it is difficult to know what role, if any, they play in producing the aphasia syndrome.

Despite these caveats, it is the case that Broca's and Wernicke's aphasia have different lesion profiles suggesting that there are functional subdivisions of neural areas within the left hemisphere. However, because the lesion data make it difficult to pinpoint a neural area 'causing' the language impairments in Broca's and Wernicke's aphasia, we will take a broad-stroke view (no pun intended) as we examine the effects of lesions on language in aphasia. We will still map language to neural areas of the brain, but we will focus on neural systems associated with the components of language and not associate a specific language function to one 'place'.

## 2.5    The Components of Language: Putting Language Together

Consider the pieces of language that you need when speaking or listening. You need to be able to articulate and perceive the speech sounds of English (or whatever language you may know); these sounds need to be put together to form the different words that make up your mental dictionary (lexicon); the words need to be put together into sentences; and the words and sentences need to convey some meaning. Figure 2.4 shows these components. As you can see, they are organized hierarchically – for understanding, we go from sounds to words, and for speaking, we go from meanings (ideas) to sounds.

What do we do with this information? Is this the end of the story? Not at all. Figure 2.4 is a start since it identifies the individual components comprising language. But naming each component and describing it as a box, separate from the other components, fails to say what is going on <u>inside</u> each box and how that box communicates with the other boxes. To do so requires building a model or framework. The framework we adopt here is broadly-based on connectionist models (Rumelhart et al. 1986) which use simplified properties of the brain to characterize the architecture of cognitive systems. For us, this framework provides a critical link between language and the brain.

There are a number of properties of the connectionist framework that will inform our discussion of the effects of brain injury on language. Consider what we know about neurons. Neurons connect to other neurons. How they 'fire', meaning how they are activated or inhibited, is determined by the strength of the connections to other neurons. Thus, one neuron does not do one thing. It is part of a network.

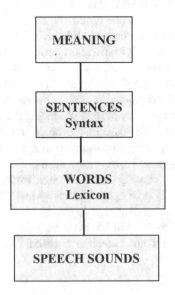

**Fig. 2.4** The components of language

The same principle holds for language and its components. Here, a sound or a word or a meaning does not reside in a single node (a node is analogous to a simplified neuron). Rather, as Fig. 2.5 shows, the word 'cat' is represented by the interaction of a large number of individual nodes. It is the *pattern* of activation of this network of nodes, not a single node that represents the word 'cat'. This means that there is not a 'cat' cell, a 'house' cell, or a 'grandmother' cell.

A sound, a word, or a meaning then is 'defined' in terms of a network of nodes that are connected to each other. These nodes interact with each other via patterns of activation and inhibition, much like the network of neurons in our brain. It is the pattern of activation of the network that 'defines' individual words, sounds, and meanings.

As we will see in the chapters that follow, each component depicted in Fig. 2.4 has its own organization or architecture. Critically, each component of language is organized in a network-like architecture. That is, there is a network of sounds, a network of words, and a network of meanings. What this means is that an individual sound such as 'd', an individual word such as 'cat', or an individual meaning such as 'animal' is not a silo, independent and separate from other words, sounds or meanings. Rather, it is part of a larger network. Thus, the sound 'd' is connected to other sounds in English, the word 'cat' is connected to other words, and the meaning 'animal' is connected to other meanings. The nature of the connections within each component affects not only how a particular sound, word or meaning is accessed, but it also affects in important ways the words, sounds, and meanings connected to it.

Further details of network connections in each component of language will be discussed in Chaps. 3, 4 and 5. However, to provide a brief introduction of what we mean by a network and its organization, let's consider a very simplified lexical network with the 8 words shown in Fig. 2.6. We know thousands of words – estimates

**Auditory Input**                                                                         **Cat**

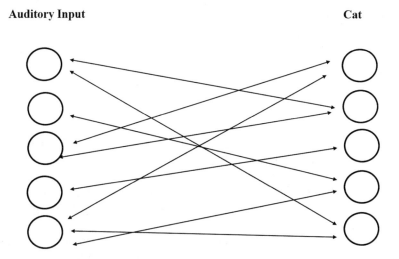

**Fig. 2.5** The pattern of activation of the nodes (simplified neurons) representing the word 'cat'. Each circle (on the left) corresponds to a node activated during auditory input. This auditory input is connected to and activates or inhibits the nodes which together represent the word 'cat'

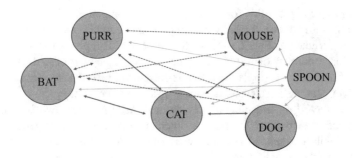

**Fig. 2.6** A simplified lexical network. Connections between words in the network vary in their connection strengths. Red lines represent strong connections, red dashed lines weaker connections, and green lines the weakest connections

range from 20,000 to 40,000 words. Based on these numbers, the ability to retrieve a word we want to use would seem daunting. And yet, we can come up with the word we want typically in a matter of milliseconds. How words are represented may make our task easier. For the purposes of this example, the network in Fig. 2.6 is the totality of the words in English. Here, there are connections with words that sound similar such as 'cat' and 'bat'; there are connections based on descriptive attributes such as 'cat' and 'purr'; and there are connections between words based on associations with each other such as 'cat' and 'dog' and 'cat' and 'mouse'.

Critically, as shown by the different arrow types, there are differences in the *strengths* of these connections as well. There are strong connections between words that sound similar, 'cat' and 'bat', that share descriptive attributes, 'cat' and 'purr', and that are associated with each other, 'cat' and 'dog' and 'cat' and 'mouse'. There are weaker connections between 'dog' and 'purr', 'mouse', and 'bat' because these words are only indirectly related via the connections 'dog' has with 'cat'. The word 'spoon' is minimally connected to the other words in this network because it shares no properties with any other word. Nonetheless, it does have a weak connection to the other words because it is a word and hence is a part of the mental lexicon.

Of importance, the nature of the connection strength affects how words are 'activated' in the network. Say you hear the word 'dog' (see Fig. 2.6). 'Dog' will be the most strongly activated because it matches the auditory input you just received. There will be different degrees of activation of the other words in the network depending on the strength of their connections with 'dog' and with other words in the network. This means that the activation of words in the lexicon is *graded*; given 'dog' as input, 'cat' will be partially activated because it is strongly connected to 'dog'. 'Purr' will be weakly activated because, despite the fact that dogs don't purr, 'purr' is connected to 'cat' which is strongly connected to 'dog'. 'Spoon' will have the weakest activation.

Let's think about what this means. If words and sounds and meanings are each connected in a network, then when you and I are speaking or listening, the networks associated with each component are activated. Remember, the slips of the tongue we talked about in Chap. 1? The nature of the network connections provides an 'explanation' for why you might have said 'aunt' for 'mother' and not 'shoe' for 'mother'. Even though 'mother' was the target word, it is strongly connected in the network by meaning to 'aunt' but only weakly connected to 'shoe'. Because it is strongly connected to 'mother', 'aunt' will be partially activated, even if you are thinking about your mother. In fact, because of their close connection in the network, the two words, 'aunt' and 'mother', *compete* with each other, and your system has to *select* the word you will ultimately use. In the case of the slip of the tongue, the competitor word 'aunt' was selected rather than 'mother', and out came the wrong word. Similarly, given the functional architecture of the network in Fig. 2.6, you might say 'cat' for 'dog' but you would not mistakenly say 'spoon' for 'dog' because the connection strengths between 'spoon' and 'dog' are so weak.

In sum, the components of language (sounds, words, meanings) are characterized in terms of networks of connections between nodes just as neurons in the brain operate as networks to shape behavior. Critically, the components themselves are not isolated from each other. As shown in Fig. 2.4, they are connected to each other so that information from one component can flow to another component. Sounds need to form words; words are put together into sentences; and both words and sentences have meanings.

## 2.5.1  Interactivity: Information Flow in the Network

In connectionist models, information flow goes in both directions – it goes from sounds to meanings and meanings to sounds, and the amount of activity in one component influences the amount of activity in another component. This means that the system is *interactive*. For example, consider information flow from the component for sounds to the component for words. Let's say you are chewing bubble gum while you are attempting to say the word 'pear' and it affects the clarity of your production. Distorted productions influence how you articulate sounds affecting the ultimate acoustic patterns of your speech output. Thus, because the sounds 'p', 'ea', 'r' are distorted, they will be more weakly activated compared to their activation were you to talk clearly without chewing gum. The degree of activation of the individual sounds in the speech component will in turn have a cascading effect on the other language components. The word 'pear', in the lexical component will be less strongly activated as it would were the sounds produced without distortion. And because information flows throughout the language system, the weaker activation of 'pear' in the lexical component will result in weaker activation of the meaning of the word as well. Bottom line: the quality of articulation will affect the degree of activation of sounds which will influence the degree of activation of words and in turn the degree of activation of their meanings.

As this example illustrates, information flow in one component of language cascades to other components, interacting with and hence influencing the degree of activity in other components. Indeed, the interactive flow of information not only goes from sound to meaning but also goes from meaning to sound such that meaning can affect your perception of the sounds of language.

Let's say, you are talking with friends about the devastating fires in California. You again are chewing gum and do not produce your speech clearly. Your friends hear you refer to 'Smokey the ?ear', with the first sound of the last word seeming to be something between a 'b' and a 'p'. What will they think you said? 'Smokey the pear' or 'Smokey the bear'? Most likely they will hear the word 'bear', not the word 'pear'. The reason is that the *semantics* of the topic of the conversation and the *meaning* associated with the word 'Smokey' together shape the perception of the poorly produced sound. You will hear the first sound of the ambiguous word as 'b', not 'p'. 'Smokey the bear' makes sense; 'Smokey the pear' does not.

## 2.6    When Language Meets Brain

The functional architecture of language we have just described shares a number of properties with the brain. In some sense, it provides our first step in looking at how language may be represented in the brain. However, it is only a first step since we have not directly connected the functional architecture of language with that of the brain in examining the breakdown of language in aphasia. Before we do so, let's consider what might happen were we to 'lesion' the functional model of language presented in the previous sections. After all, the connectionist framework provides a model of how language is put together and its workings. The next step is to think about the effect on language when the model is damaged.

There are particular ways that language may break down given its network-like architecture. Let's first consider individual sounds, words, and meanings. Recall that it is the pattern of activation of a network of nodes and not one node that characterizes these individual 'pieces' of language. Owing to the rich number of connections of thousands of nodes corresponding to individual sounds, words, and meanings, damage to a node (or even a number of nodes) or damage to some connections between nodes will not destroy the network. Instead, it will affect the efficiency with which the particular network operates. Critically, the network architecture 'safeguards' individual sounds, words, and meanings from being lost in their entirety when there is damage to the system, preventing a *catastrophic* destruction of the system.

Nonetheless, there are 'behavioral' consequences to such damage. There will be an increase in the number of errors in selecting individual members of the network leading to what is called *graceful degradation*. It is 'graceful' because the behavior of the network is not 'all-or-none'. It is variable but systematic – sometimes coming up with correct responses, other times, not.

What types of errors will the network make? Again, the structure of the network will shape the responses. Recall that within each component of language the network-like architecture connects individual sounds, words, or meanings to each other based on the nature of their relationships and their connection strengths. As shown in Fig. 2.6, words in the network compete with each other, and the extent of this competition is based on the connection strengths between them. Damage to the network essentially introduces noise into the system affecting the efficiency with which it activates and differentiates the members of the network. Thus, although the structure of the network is maintained, the differences between words are reduced leading to an increase in competition between them. The 'behavioral' consequence is an increase in the occurrence of the types of 'slips of the tongue' we described earlier. 'Pear' might be selected instead of 'bear', 'dog' might be selected instead of 'cat', 'shoe' might be selected instead of 'foot', 'blue' might be selected instead of 'red', or 'five' might be selected instead of 'four'. 'Spoon' would not be selected for 'dog' since even in the face of damage, the two words are only weakly connected in the network. The end-result is that damage to the network results in a systematic pattern of errors; errors are not random, but rather reflect the properties of the network architecture.

## 2.7    Ready to Go

Now that we have described the clinical pictures of Broca's and Wernicke's aphasia, provided some background on the brain, and introduced a computational model of language considering potential effects of 'lesions' to the model, it is now time (perhaps high time) that we get to the main topic of this book. The material presented in this chapter provides a framework for looking at the particular effects of brain injury on language. In Chaps. 3, 4, and 5, we will dive into the effects of brain injury in aphasia on the components of language including sounds, words and their meanings, and syntax. Here, language literally meets brain. Together, their intersection provides crucial insights into the nature of language and its instantiation in the brain.

## References

Damasio, H. (1998). Neuroanatomical correlates of the aphasias. In M. T. Sarno (Ed.), *Acquired Aphasia*, 3rd ed. New York: Academic Press, pp. 43–70.
Posner, M. I., and Raichle, M. E. (1994). Images of mind. Scientific American Library/Scientific American Books.
Rumelhart, D. E., McClelland, J. L., & the PDP Research Group. (1986). Parallel distributed processing: Explorations in the microstructure of cognition. Vol I. Cambridge, MA: MIT Press.

## Readings of Interest

Goodglass, H. (1993). Understanding aphasia. Academic Press.
Joanisse, M.F. and McClelland, J.L. (2015). Connectionist perspectives on language learning, representation and processing. *Wiley Interdisciplinary Reviews: Cognitive Science*, 6 (3), 235–247. doi: https://doi.org/10.1002/wcs.1340

# What's Right and What's Wrong with Speech Sounds

**3**

## 3.1 The Sounds of Language

Why are sounds such an important part of language, and hence, a critical piece of our study of aphasia? It is because the sounds of language are the way-station to meaning. Speakers of a language share with each other a common set or inventory of sounds. These are used to designate the words in the language which ultimately map onto a meaning representation. For example, English-speakers all agree that the sounds 'd' 'o' 'g' in the word 'dog' represent the meaning of a 'dog'. In theory, our canine friend could be called a 'blik' or some other set of sounds. In this sense, the mapping between sounds, words, and their meanings is arbitrary. There is nothing intrinsic in the sounds that make up the word 'dog' that tells us that it refers to 'a furry animal with four legs, a tail, and so on'. Indeed, other languages use different sounds to represent a 'dog'; *chien* in French, *perro* in Spanish, and *hund* in German.

Speech sounds also connect us to the outside world when we speak and when we understand. A conversation cannot occur without the use of some system to convey meaning. Most of us use the auditory signal during language communication – we perceive an auditory signal when we listen and we ultimately use our articulators when we speak to produce an auditory signal that the listener perceives. Thus, we converse by mouth (for speaking) and we listen by ear (for perceiving). But there are other communication/language systems. Sign language, used by the deaf, conveys meaning by hand (for 'speaking') and by eye (for 'listening'). In this chapter, we will focus on sounds produced my mouth and perceived by ear. In Chap. 7, we will explore aphasia in those deaf individuals who use sign language as their means of communication.

As we briefly described earlier, nearly all persons with aphasia make 'slips of the tongue' producing the wrong sounds when speaking, and they make 'slips of the ear' misperceiving the sounds that they hear. This is just the beginning of the story. What we have not yet learned is *why* those with aphasia make the particular errors that they do and whether the *pattern* and *cause* of those errors differ depending on

© Springer Nature Switzerland AG 2022
S. E. Blumstein, *When Words Betray Us*,
https://doi.org/10.1007/978-3-030-95848-0_3

the type of aphasia or the area of brain injury. To understand what is going on requires looking at what those with aphasia cannot do, and also what they can do. Here, errors may provide a window into the nature of the impairments that a correct production or perception cannot. How is this possible?

Think about what it means when you converse and don't make an error. What have you learned? You have learned that all systems are 'go'; but you have learned nothing about how the pieces of language are put together or are organized. Now let's say you are invited to have dinner at your friend's house and in the midst of the conversation your friend says 'Dinner is almost ready. I just put the spaghetti in the *tot*'; 'tot'? Really? Given the context you are likely to assume that your friend meant to say 'pot'; you are less likely to consider that he was using language creatively and making up a new turn of phrase. Why? Because the sounds making up the words 'pot' and 'tot' are similar, and it makes sense given the context to assume that the intended word had been 'pot' and that 'tot' was a slip of the tongue.

Or perhaps more interesting, let's say you make an error like those made by Reverend William A. Spooner, the don of New College in Oxford, England from 1903–1924, who was famous for making sound errors in his speech. In fact, the frequency with which he made these errors gave rise to a term named after him for slips of the tongue, *spoonerisms*. It is reported that rather than saying what he intended, 'you have <u>m</u>issed my <u>h</u>istory lesson', Reverend Spooner instead said, 'you have <u>h</u>issed my <u>m</u>ystery lesson' (quoted in Fromkin 1971, p. 30). As you can see, the 'm' in 'mystery' and the 'h' in 'history' switched places. Besides being humorous (unless you were a history professor), slips of the tongue can tell us something about the processes that go into the production of speech.

Such errors are not limited to twentieth century Englanders. There were two friends conversing about their respective dogs and having a bit of an argument – instead of saying 'you have hated my dog', one of them said 'you have dated my hog'. That's a conversation stopper!

These examples of slips of the tongue tell us two things. The first is that a word is made up of individual sounds which can move. In the first example, the 'm' of 'missed' and the 'h' of 'history' changed places. Similarly, in the second example, the 'h' in 'hated' and the 'd' in 'dog' changed places. That sounds can move from one word to another tells us that *words are not indissoluble wholes*. Rather, a word is comprised of 'individual' sounds that have to be assembled or put together when speaking.

The second thing that the two slips of the tongue tell us is that in order for the sounds to change places, we have to *pre-plan* the words in the sentence we want to say before we say them. If not, how could a sound occurring in a later word be moved to an earlier word unless it was already in the 'mind' of the speaker. This type of error shows that when we produce phrases or sentences, we do not speak one word at a time, selecting each word separately. Instead, we plan larger spans of words as we are talking (see Sect. 3.2.4 later in this chapter and Chap. 5 where we will discuss how words are put together into sentences when we talk about syntax).

Analyzing just these few examples of slips of the tongue has told us a lot about how we produce sentences. Now consider what we can learn when we look at the effects of brain injury on speaking and perceiving where word betrayals are common. We will see that errors or deviations from the norm are not only informative, but provide answers to questions that would not have been possible had we simply looked at production and perception errors in non-brain-injured individuals. Do sounds reside in one place in the brain? What aspects of speech are affected by brain injury and what aspects are preserved? Do sound errors in production or perception mean that the intended sound is lost, gone forever? Does a failure to perceive sounds correctly result in an inability to understand language – after all, how can a word or a sentence be understood if the sounds of language are misperceived? These are only some of the questions we will be asking as we now turn to the effects of brain injury on speech production and speech perception in aphasia.

## 3.2 Saying What You Want to Say: Speech Production

All aphasics make speech production errors. They produce a variety of types of errors including the substitution of one speech sound in a word for a different one, the omission or addition of a sound in a word, or the misordering of the sounds either within a word or between words. Table 3.1 shows examples of each of these types of errors.

Of importance, the same type of errors as those shown in Table 3.1 occur irrespective of clinical syndrome or lesion site. That is, those with Broca's aphasia and frontal lobe lesions and those with Wernicke's aphasia and temporal lobe lesions produce similar errors. These findings provide the first suggestion that there is not one area of the brain dedicated to speech production. Rather, the speech system encompasses multiple brain areas of the frontal and temporal lobes that support the production of speech. What we don't know at this point is what the nature of this system is – what is its architecture and how is it affected by brain injury. It is to this that we now turn.

### 3.2.1 The Speech Network

Recall that in Chap. 2 we indicated that each component of language is represented in a network-like structure. There we described how each word is characterized in terms of patterns of activation of a network of nodes. We described a simplified

**Table 3.1** Speech production error types

|  | Target | Error |
|---|---|---|
| Sound substitution | time | kime |
| Simplification | green | geen |
| Addition | saw | staw |
| Order of sounds in a word | degree | gedree |
| Order of sounds between words | read books | bead rooks |

lexical network (shown in Fig. 2.6) where the connection strengths between words were a function of how similar words were to each other in sound and/or meaning. The same is true for sounds. The more similar sounds are to each other, the greater the connection strength between them. What do we mean by similar sounds?

Similarity is based on how close speech sounds are to each other in terms of how they are produced and what their acoustic realizations are. To produce the speech sounds of language, we make a range of articulatory gestures. These gestures reflect the different articulatory movements made by the speech apparatus as a sound is produced including how the sound passes through our speech apparatus (its *manner of articulation*), what position our tongue is in when we are producing the sound (its *place of articulation*), and what our vocal cords are doing (its *voicing characteristics*). These different articulatory gestures give rise to corresponding acoustic patterns. Thus, changes in articulatory gestures and their acoustic patterns result in the different sounds of language.

Just as we learned in the early part of this chapter that words can be broken down into individual sounds, so sounds can be further broken down into parts or features, each feature having an articulatory and acoustic correlate. In this way, every speech sound in language is comprised of a set or bundle of articulatory/acoustic features that uniquely define it and differentiate it from other speech sounds. It is not necessary that you know the fine details of these properties. But what is important is to understand that these parts or features are the building blocks of the sounds of all languages.

Figure 3.1 shows a simplified network for the speech component. The bottom row indicates the articulatory gestures corresponding to manner of articulation, place of articulation, and voicing. Each of these gestures can be realized by different features shown here in acoustic and articulatory terms. As listeners, we have to perceive the acoustic features of speech and as speakers we have to produce the articulatory features of speech. For manner of articulation, speech sounds may be produced with a closure (the feature stop) or a constriction (fric: fricative) in the mouth. For place of articulation, the tongue may be at the lips (lab: labial), behind the teeth (alv: alveolar), or at the back of the mouth (vel: velar). And for voicing, the vocal cords can be held apart (vl: voiceless) or vibrate (vd: voiced).

These features uniquely combine to define individual speech sounds. Thus, the sounds 'z', 'd', 'b', and 'p', shown in Fig. 3.1, are each realized by a set of features, one corresponding to manner of articulation, one to place of articulation, and one to voicing. It is the relationship between these sounds that defines how similar they are to each other. As you can see, 'z' and 'd' share all features except manner of articulation; 'z' is a fricative and 'd' is a stop. 'd' and 'b' share all features except place of articulation; 'd' is alveolar whereas 'b' is labial. And 'b' and 'p' share all features except voicing; 'b' is voiced and 'p' is voiceless. Thus, each of these pairs of sounds is distinguished by only one feature. In contrast, 'z' and 'b' share the feature voice and are distinguished by two features, manner of articulation ('z' is a fricative and 'b' is a stop) and place of articulation ('z' is alveolar and 'b' is labial). 'z' and 'p' are distinguished by 3 features contrasting in manner of articulation, place of articulation, and voicing.

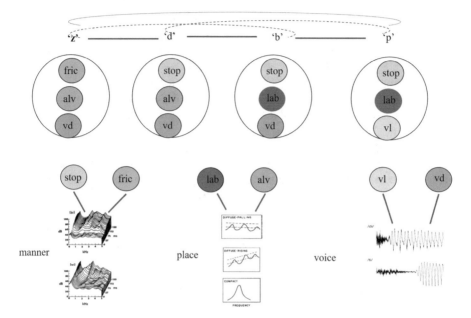

**Fig. 3.1** A simplified speech network. The articulatory and acoustic features that define the sounds in the network are shown in the bottom row. The manner, place, and voicing features that are associated with each of the sounds 'z', 'd', 'b', and 'p' are shown together in the next row up. The relationship between these sounds is shown with the strongest connections in the network indicated by a dark solid red line for sounds distinguished by one feature, weaker connections in the network indicated by the dotted line for sounds distinguished by two features, and the weakest connections indicated by the solid light red line for sounds distinguished by three features. Bottom left figure reprinted with permission from Mack, M. and Blumstein, S. E. (1983). Further evidence of acoustic invariance in speech production: The stop–glide contrast. The Journal of the Acoustical Society of America, 73(5), 1739–1750. Copyright 1983, Acoustic Society of America. Bottom middle figure reprinted with permission from Blumstein, S. E. and Stevens, K. N. (1979). Acoustic invariance in speech production: Evidence from measurements of the spectral characteristics of stop consonants. The Journal of the Acoustical Society of America, 66(4), 1001–1017. Copyright 1979, Acoustic Society of America

Because of the network architecture, features and sounds are not isolated from each other but are connected to each other. Thus, the activation of one sound and its features will partially activate other sounds in the network, and the degree of their activation will be a function of how similar they are to each other. The greater the similarity, the greater will be the strength of the connections between them and the more likely they are to influence each other. As shown by the different red lines connecting the sounds in Fig. 3.1, the strongest connections are between sounds that share all acoustic-articulatory features but one. There are systematically weaker connections between sounds as a function of the number of features they share and the number of features that distinguish them.

Let's consider the behavioral consequences of such a network architecture. Let's say you are in a conversation with a friend and are preparing to say the 'p' in the word 'pen' in the sentence, 'I said pen'. You would select the word 'pen' from your mental dictionary including its sounds and the features comprising them and ultimately articulate the 'p', producing the acoustic patterns that will be perceived by your friend. Not only will the 'p' be activated but other sounds in the network will also be partially activated. However, it turns out that as your conversation goes on, you are distracted and you do not produce 'p' but rather you produce a different sound; you have made a slip of the tongue. What is your best guess about what sound or sounds you might produce in error?

The answer is you would most likely produce a similar sound, one that shares articulatory and acoustic properties with 'p'. Thus, you might say 'ben', a voicing error, but you would less likely say 'zen' since 'z' shares few features with 'p'; the two sounds differ by three features: manner, place and voicing (if you are not sure check Fig. 3.1).

What about aphasia? What is your best guess about what happens when the wrong sound is produced resulting in the substitution of one sound for another?

### 3.2.2   Where Sound Substitution Errors Come from

Sound substitutions are the most common error type produced in both Broca's and Wernicke's aphasia. The first question to ask is when aphasics substitute one sound for another, does it mean that they literally have 'lost' that sound and can no longer produce it, forever now using the substituted sound. In other words, looking at the example in Table 3.1 where 'kime' was produced instead of 'time', is 'time' always produced as 'kime' and are all words that have 't' in them now produced with a 'k'? In such a case, the sentence 'Tom put tomatoes on his toast' would be produced 'kom puk komakoes on his koask'. Does this happen? Absolutely not! In aphasia, sounds are not lost; rather they are *substituted* for each other. Sometimes a 't' is produced as a 'k'. Sometimes it is produced correctly. And sometimes it is produced as another sound that is similar to it. This pattern turns out to be the case irrespective of clinical type of aphasia.

The architecture of the network provides an explanation for why speech sounds are not lost. Just as we discussed in Chap. 2, a word is not represented by a single node, but rather is represented by a network of nodes (see Fig. 2.6). The same is true for a speech sound and the features comprising it. In this way, brain injury to the network will not destroy individual sounds or their features but may affect the *efficiency* of the network in accessing and articulating the sounds of language.

That sounds are not lost may disappoint those of you who are Steve Martin fans and saw the movie 'The Man with Two Brains'. As I recall there is a scenario in this movie that is particularly relevant to current theories about language functions of the brain, and so we will take a brief detour and describe the events of relevance. (In truth, my recollection may be imperfect as I have not been able to find the exact scenario in the DVD version of the movie. What I recall from the movie which I saw in 1983 may reflect an extension of my own imagination of a scene that is in the movie. So forgive me if I take poetic license and give you my own rendition that is in the 'spirit' of the movie).

Imagine the following scenario. Without lots of detail, Steve Martin has fallen in love with a woman whose brain has been excised from her skull and placed in a glass translucent container with a nutrient solution that has kept her alive, perfectly sentient, and able to speak and hear. Steve Martin kidnaps her. He runs away to a rented cottage where he places the container holding her brain on a table. He then leaves the room. The sun is pouring in and a single ray of light hits one small section of her brain. Steve Martin returns to the room and sees that area of the brain smoking. He becomes very upset. Panicked that the heat of the sun has damaged that spot on the brain, he conducts a brief clinical examination. First, he asks his lover to recite the alphabet. She does this but stops at 'y', not saying 'z'. He becomes agitated and concerned. He then asks her a series of questions: 'what is the black and white striped animal?' She replies 'an ebra'. 'Where does it live?' She replies, 'in the oo'. 'Oh no, oh no', he cries in despair. 'Her z's are gone'.

This is a perfect example of an extreme version of modern-day *phrenology*, a theory originally proposed by Franz Gall (1759–1828), who claimed that there was a one-to-one relation between areas of the brain (identified through bumps on the head) and faculties of mind (thoughtfulness, kindness, and so on). The basic premise of this theory is still alive today embedded in the theory of modularity proposed by Jerry Fodor in 1983 in his monograph *The Modularity of Mind*. In Fodor's view, the mind is divisible into separate modules associated with cognitive faculties, language being one of them. Modules have a number of properties, among them, a 'fixed' neural architecture (Fodor 1983, p. 98). Taken further, some have proposed that the components of language we talked about in Chap. 2 are localized in distinct neural areas: speech perception located in the superior temporal gyrus of the temporal lobe, speech production in areas within the frontal lobe, words in the parietal lobe, meanings in the middle temporal gyrus in the temporal lobe, and syntax in the inferior frontal gyrus in the frontal lobe.

Returning to Steve Martin. Steve Martin's love, who lost her 'z's', showed an impairment which reflected a literal view of the modularity of language with individual sounds represented in a specific area of the brain. Injure that one spot and out go individual sounds. Well, the facts of aphasia do not support such a view. Sound substitution errors indicate that one sound is simply substituted for another. As for speech sounds themselves and those that result in sound substitutions, they do not have a 'fixed' neural architecture located in one place in the brain. Rather, substitution errors occur with damage to multiple structures located in the frontal as well as temporal areas of the brain.

If different sounds can substitute for each other, then the next question to ask is whether there is a systematic *pattern* associated with such substitutions or whether substitution errors are random, such that any speech sound can serve as a substitute for any other. And is the pattern of substitutions different across clinical syndromes and neural loci?

Given our description of the speech network, we would expect more substitution errors to occur between similar sounds. And that is the case. The closer the sounds are in the network based on their articulatory and acoustic similarity, the more likely they are to undergo substitutions. Thus, there are more errors between sounds distinguished by one feature than sounds distinguished by two or more features.

Critically, this error pattern emerges in both Broca's and Wernicke's aphasia, providing more evidence that the speech system is distributed involving both anterior and posterior brain structures.

These findings provide another piece of information on the effects of brain injury on language and, in this case, on speech production. Namely, the architecture of the network is preserved irrespective of the clinical type of aphasia and neural locus of the lesion – speech sounds are retained and the pattern of errors reflects the way the sounds of speech are represented and organized.

Finding that the structure of the network is preserved is crucial in understanding what is 'right' or retained when someone has aphasia. However, as just shown, we also need to identify what is 'wrong' or impaired. What is the cause of the speech production errors? What conditions would give rise to the pattern of substitution errors?

The deficit lies in how the network is activated. Let's think about it. Consider what would happen if, because of brain injury, the system is functioning as though in noise (see Fig. 3.2). As you can see, the basic structure of the network remains. However, the noise in the system affects the precision with which the network operates, leading to a reduction in the overall activation of sounds and their features as well as a reduction in the differences between them. Because distinctions between sounds and features are less clear-cut, the more similar sounds are to each other, the

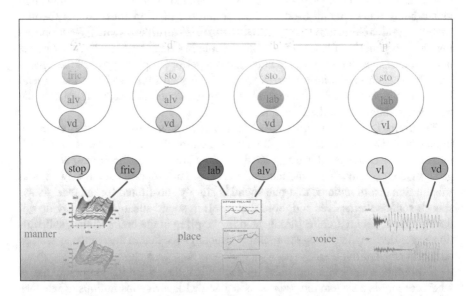

**Fig. 3.2** Effects of noise in the speech network. The introduction of noise leaves the structure of the network intact, but reduces the degree of activation of sounds and their features as well as the differences between them. Bottom left figure reprinted with permission from Mack, M. and Blumstein, S. E. (1983). Further evidence of acoustic invariance in speech production: The stop–glide contrast. The Journal of the Acoustical Society of America, 73(5), 1739–1750. Copyright 1983, Acoustic Society of America. Bottom middle figure reprinted with permission from Blumstein, S. E. and Stevens, K. N. (1979). Acoustic invariance in speech production: Evidence from measurements of the spectral characteristics of stop consonants. The Journal of the Acoustical Society of America, 66(4), 1001–1017. Copyright 1979, Acoustic Society of America

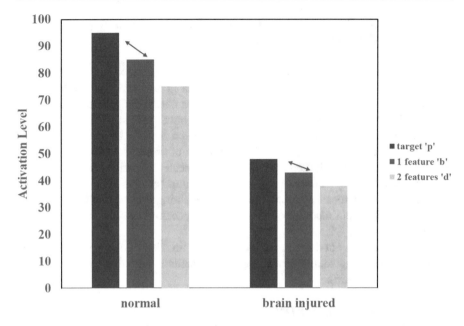

**Fig. 3.3** Activation levels in the speech component under normal conditions and when brain-injured. Activation of the target sound 'p' is compared to 'b' and 'd' under normal conditions and when there is noise in the speech system due to brain injury. Note that noise reduces the distance between 'p' and 'b' (shown by the red arrows) when the system is brain-injured compared to when it is normal. See text

less easily will their differences be maintained, and the more likely will similar sounds be substituted for each other.

Let's consider the effects of noise on the speech network in more detail. Assume that under normal circumstances, the representation associated with each sound of the language, in this case English, typically excites the network at an activation level of 95%. Say the speaker wants to produce 'p'. The activation of 'p' will be at 95%. Similar sounds will also be partially activated. For the purposes of this example, when 'p' is the target sound, 'b', distinguished from 'p' by one feature, will have a lower activation level, say 85%, and 'd', distinguished from 'p' by two features, will be activated at say a 75% activation level. Let's now assume that because of brain injury, there is a 50% reduction in the degree of activation of speech sounds. As Fig. 3.3 shows, the introduction of this noise in the system has two consequences. First, it reduces the activation of the target sound. Second, it reduces the *difference* between the target sound and those sounds that share features with it, rendering the activation patterns of the sounds distinguished by one feature to be less distinct. Thus, noise in the system not only reduces the efficiency of accessing the appropriate sound, but it also reduces the magnitude of the difference between similar sounds. The result? There will be increased variability in producing the correct sound, and an increased probability that a similar sound will be substituted for the target sound.

### 3.2.3   Speech Production Differences Between Broca's and Wernicke's Aphasia

Up to now, we have not seen any differences in speech production that distinguish Broca's and Wernicke's aphasia. Can this be correct? The description of the clinical characteristics of these two syndromes described in Chap. 2 suggested clear differences in the character of speech production. What have we missed?

There is one clear difference that emerged when looking at the speech errors described in Table 3.1. Despite the similar pattern of errors that occurred, the raw number of errors differed; there are many more speech production errors in Broca's aphasia compared to Wernicke's aphasia. These numbers are consistent with the clinical picture of Broca's aphasia. The major presenting symptom of Broca's aphasia is a speech output disorder often producing speech that to the listener sounds distorted.

Perhaps, we have not dug deep enough into the fine details of speech production in order to identify potential differences across patients with distinct clinical syndromes and lesion sites.

#### 3.2.3.1 The Devil is in the Details

Although we may perceive distortions in the speech output of those with Broca's aphasia, our ears cannot easily evaluate what may be the nature of such articulatory impairments. Doing what are called 'close transcriptions' of speech by trained linguists and particularly transcriptions of those who have sustained brain injury is fraught with difficulty. Native speakers' perception of the sounds of their language are shaped by the language and the dialect they know and speak, and being able to reliably transcribe fine distinctions or distorted productions is at best a challenge. In addition, speakers do not have sufficient articulatory control to say exactly the same thing every time. Thus, it is difficult, if not impossible, to determine what articulatory gestures underlie pathological productions by simply listening. In order to make such a determination requires looking at the fine details of the speech output. Acoustic analyses can provide such a window.

Examination of the acoustic patterns of a variety of speech sounds has shown that those with Broca's aphasia typically have articulatory impairments. For example, looking back at Fig. 3.1, you can see that the feature description of 'p' differs from that of 'b' by one feature; 'p' is voiceless and 'b' is voiced. The articulatory difference between voiced and voiceless consonants is the coordination between what the articulators in the mouth are doing and the state of the vocal cords (the larynx). That is, it is the *timing* between the release of the consonant and the vibration of the vocal cords that determines whether a consonant like 'p' is voiceless and a consonant like 'b' is voiced. In the case of 'p', there is a delay in vocal cord vibration relative to when the lips release the stop closure, whereas in the case of 'b' the vocal cords start to vibrate about the same time as the release of the stop closure. This acoustic measure is called *voice-onset time (VOT)*.

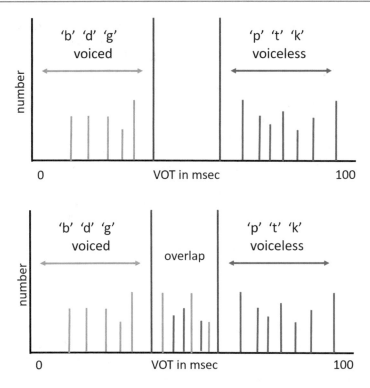

**Fig. 3.4** Voice-onset time productions of a typical speaker without brain injury (top panel) and someone diagnosed with Broca's aphasia (bottom panel). The top panel shows the range of VOT values for voiced ('b', 'd', 'g') and voiceless ('p', 't', 'k') stop consonants produced by a non-brain injured speaker. Voiced stops (in blue) have short VOT values, and voiceless stops (in purple) have long VOT values. Note that there is a separation between the VOT values for voiced and voiceless stops. The bottom panel shows the VOT values produced by a Broca's aphasic. In contrast to the productions of non-brain injured speakers, there is overlap in the VOT values for voiced and voiceless stop consonants

It is possible to measure VOT by analyzing the time in milliseconds between these two articulatory events. Voiceless stops like 'p' have long VOTs and voiced stops like 'b' have short VOTs. The top panel of Fig. 3.4 shows an example of the range of VOT values that define voiceless stops and the range of VOT values that define voiced stops.

Voice-onset time analysis of productions of Broca's aphasics indicates that they have impairments in coordinating the timing of their vocal cords with the release of the stop closure. As shown in the bottom panel of Fig. 3.4, many productions show overlap in voice-onset time values between the areas defining voiced and voiceless sounds. Such distortions do not occur in Wernicke's aphasia.

The presence of articulatory deficits in Broca's aphasia is not surprising given that their lesions often extend to areas involved in motor programming and articulatory implementation of speech. These areas include the premotor and motor cortex, insular cortex and the basal ganglia (see Chap. 2 and Fig. 2.3). Critically, we have identified an important difference between the speech production of Broca's and Wernicke's aphasia, one determined by a distinct but neurally distributed function for articulation in frontal areas of the brain.

Another characteristic of the speech output that reflects the functional specialization of frontal areas in speech articulation is the fluency of speech output. As we have described, it is common for the production of speech in Broca's aphasia to be non-fluent. That is, output is slow, labored, and there are frequent pauses between words. This non-fluency reflects difficulty in initiating and actually producing the sounds of speech. In contrast, the speech production of those with lesions in posterior areas including Wernicke's aphasia is fluent. Not only is speech well-articulated, but it comes out easily with few pauses, if any, between words.

### 3.2.4   Planning Ahead Before We Speak

We have learned a lot about the production of sounds in aphasia. But to this point, we have looked in detail only at the production of individual sounds. As we described earlier, the occasional slip of the tongue made by you and me show that we plan ahead as we speak. It is important to know whether brain injury in aphasia affects this ability. Imagine being unable to plan more than a sound at a time or a word at a time. The effects on speaking and ultimately communicating would be devastating.

Although we may think that when we produce sounds, we say them one at a time, as though the sequence of sounds that make a word are put together when we articulate them like 'beads on a string'. This is not the case. Rather, we produce a series of articulatory movements that connect the sounds together seamlessly. In fact, how we produce a particular sound is influenced by its context – what comes before it and what comes after. This requires planning; you have to know what you want to say before you say it.

It turns out that planning ahead also occurs in aphasia. As Table 3.1 showed, sounds can change places within a word and also between words. For example, 'degree' was pronounced as 'gedree'. Here, the 'd' and 'g' changed places and the later sound, 'g', was incorrectly produced at the beginning of the word. For this to happen, the whole word 'degree' had to be pre-planned before it was produced.

Such exchange errors also occur between words in a phrase or a sentence. For example, the order of the initial sounds in the phrase 'read books' was realized as 'bead rooks'. This means that the words were not accessed and then produced one at a time; rather, both words were planned prior to their production.

## 3.3 Listen to Me: Speech by Ear

Imagine the different places where you may have a conversation with a friend – you could talk in the library, on the street, in a restaurant, or during a live concert of your favorite rock band "Queen". As you might expect, your ability to understand what is being said to you will become increasingly difficult because the ambient noise will mask out the speech signal. Thus, even under normal circumstances, speech occurs in a noisy environment which can make it a challenge to follow what is being said.

There are other challenges you and I face in trying to understand a spoken message. This is because, as we have discussed earlier, the sounds of speech are not produced one at a time. Rather, they are produced in the context of other sounds, and this context exerts a considerable influence on the acoustic manifestation of that sound. Let's compare the production of the syllables 'su' and 'si'. When you produce 's', your lips are rounded in the context of the vowel 'u' but not for 'i'. Sure, you say (I hope). So what's the problem? You still heard the first sound as 's' in both cases – a slam dunk. In actuality, as Fig. 3.5 shows, the acoustic signal associated

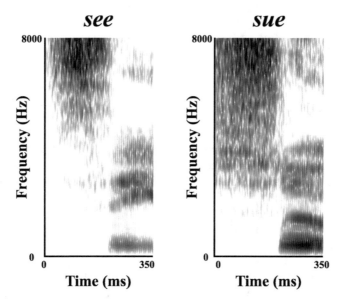

**Fig. 3.5** Spectrograms showing a time by frequency plot of a typical speaker's production of 'see' and 'sue'. In each, there is frication noise at the first part of the production reflecting the narrow constriction for 's' followed by formants (resonant frequencies of the vocal tract) associated with the vowels 'i' in the case of 'see' and 'u' in the case of 'sue'. Note that the frequency of the noise differs in the two productions; the 's' in 'sue' extends to lower frequencies than the 's' in 'see' because the lips are rounded in anticipation of the vowel 'u' which is produced with lip rounding. Rounding the lips extends the length of the vocal tract lowering the frequency of the noise and the second formant in the vowel

with 's' in 'see' is not the same as the 's' in 'sue'. How does your brain then 'know' that the two different acoustic patterns associated with 's' are the same sound?

What this example shows is that there is not a simple one-to-one relationship between the information in the acoustic signal and the sounds of speech. Figuring out how to unpack the complex acoustic signal in perceiving the sounds of language is, as yet, an unsolved problem, and it is beyond the scope of this book to detail the various proposals on how this is done. But the bottom line message is that to the listener, the sounds of language seem to be stable and fixed. However, in reality, the sounds of language are complex and varying. Consider what this means for someone with aphasia. Depending on the area of brain injury, one might expect deficits in perceiving the sounds of speech. Such deficits could have a devastating effect on understanding words and on language comprehension more generally. After all, if speech is the 'gateway' to meaning, misperceiving sounds should result in failures to understand.

In the following section we will examine whether the perception of speech is damaged in aphasia. Similar to our discussion of speech production, we will focus on those aspects of speech perception that are impaired and those aspects of speech perception that are spared. Together, they will provide answers to a number of questions. We will be able to determine whether the functional architecture of the speech perception apparatus is impaired or not; whether speech perception deficits are the 'cause' of failures to understand language; whether the speech perception system is 'localized' or 'distributed'; and if the latter, whether different neural areas have different functions in speech perception.

### 3.3.1   Perceiving the Differences Between Sounds of Language

We rarely think of sounds of language as separate from words. But words have meanings. So were we to simply test whether persons with aphasia have difficulty perceiving the differences between words like 'tower' and 'power', we would be unable to determine whether an incorrect response is due to a failure to perceive the difference between the sounds 't' and 'p' or the difference in meaning between the words 'tower' and 'power'.

One way to examine this issue is to set up a test to see how well persons with aphasia can determine the differences between sounds that make up a *nonword*. (A nonword is a possible but non-existent word in English; thus, it has no meaning). For example, the nonwords 'pem' vs. 'bem' share the same sound contrast, 'p' vs. 'b', as the words 'pet' vs. 'bet'. Presenting to those with aphasia pairs of nonwords for judgments about whether they are the same, 'pet' vs. 'pet', or different 'pet' vs, 'bet', a clear pattern of results emerges. All patients show difficulties, but the degree of difficulty varies as a function of the type of aphasia and lesion site. Those with lesions involving the superior temporal gyrus or presenting with Wernicke's aphasia show severe deficits in discriminating speech sounds, whereas those with lesions involving the inferior frontal gyrus or presenting with Broca's aphasia show deficits, but they are mild. Considering the neuroanatomy of the brain, it is not surprising that damage to the superior temporal gyrus would result in poor performance in discriminating the sounds of language. After all, the primary auditory areas are located in the temporal lobe and are involved in the processing of auditory information including speech.

The number of errors informs us about the severity of the speech perception deficit. However, it does not tell us anything about the pattern of errors that occur. Looking back at the architecture of the speech component shown in Fig. 3.1, we saw that the sounds of language are broken down into smaller components or features. Similarities between sounds can be measured in terms of the number of features they share and the number of features that distinguish them. As you might expect, the fewer differences between two sounds, the harder it will be to distinguish them.

Let's try it. Listen in your mind's ear to the following pairs of sounds – 'ba' vs. 'pa' and 'ba' vs. 'sa'. Which pair do you think is more similar? The answer is 'ba' vs. 'pa'. In terms of features then, the more similar sounds are to each other, the harder they will be to discriminate from each other. Thus, sounds distinguished by one feature are harder to discriminate than sounds distinguished by more than one feature (see Fig. 3.1). This is what happens when you and I are asked to discriminate speech in background noise.

Exactly the same pattern emerges when persons with aphasia are asked to discriminate speech sounds without any accompanying background noise. Importantly, however, similar to what we saw in the previous section on speech production, there is variability in performance; sometimes errors occur in discriminating a particular pair of stimuli, and other times the stimulus pair is correctly discriminated. These findings indicate that the feature contrast is not lost, but rather is vulnerable to error after brain injury.

Importantly, this pattern of breakdown is the same for those with Broca's aphasia who have frontal lesions and those with Wernicke's aphasia who have temporal lesions. These findings mirror the patterns we saw in the previous section on the production of speech in aphasia, and the explanation is fundamentally the same. The speech component is neurally distributed encompassing temporal and frontal brain structures. Critically, the architecture of the speech component is preserved in aphasia. Brain injury introduces 'noise' into the speech network resulting in speech perception impairments characterized by more errors in discriminating sounds as a function of how similar they are to each other.

That may be all well and good, but how do we know that the deficit involves the processing of speech sounds and not discriminating words based on their meanings? We can answer this question by comparing how persons with aphasia fare when asked to discriminate sounds in nonwords compared to when they are asked to discriminate sounds in words. Consider that when listening to nonwords, all the listener has to go on are the differences between two sounds, whereas when listening to words, the listener can attend not only to the differences between the two sounds in the words but also the differences between the meanings of the two words. I expect that your predictions are that it would be easier to discriminate words that sound similar than to discriminate nonwords that sound similar because you have more information available to you. Right?

Let's check this out with an example. For 'pem' vs. 'bem' all you have to go on is the difference in the initial sounds. In contrast, you can distinguish 'pet' vs. 'bet' not only by sound, 'p' is different from 'b', but also by meaning; 'Rover' is not the same as 'a wager'.

This is exactly what the results show in aphasia. It is considerably more difficult for them to discriminate speech sounds in nonwords than in real words. When those with aphasia cannot depend on meaning to discriminate similar sounding inputs, their performance plummets. This means that both those with Wernicke's and Broca's aphasia have deficits in perceiving the sounds of speech.

### 3.3.2  Can You Understand if You Have a Speech Perception Deficit?

Although both those with Wernicke's and Broca's aphasia have problems in discriminating the sounds of speech, those with Wernicke's aphasia have the more severe deficit. In addition, we know from talking to those with Wernicke's aphasia and clinically evaluating them that they have severe auditory language comprehension deficits as well. Is it possible that these two deficits are linked? That is, could a failure to process the sounds of speech underlie the ability to understand language?

Let's say I present to someone with Wernicke's aphasia, a card with four pictures on it: 'pot', cot', 'dot', and 'tot'. The task is to point to the 'pot'. What if the 'p' in 'pot' is misperceived as 'k' (the first sound of 'cot')? The patient will point incorrectly to the picture of the 'cot'. So on the surface, it would appear that there is a comprehension failure; the Wernicke aphasic does not understand the word 'pot'. However, there is another possibility. Namely, the word 'pot' and all the words in the stimulus array may in fact be understood. However, 'p' is misperceived as 'k' resulting in pointing to 'cot' instead of 'pot' in error.

Now consider a more extreme example. What if a person with Wernicke's aphasia were given a sentence like 'the boy likes his ham on toast' followed by a question about what the boy likes to eat. What if the sentence is misperceived as 'the poy rites his kam om poask' – the Wernicke would undoubtedly fail to answer correctly, and you would be convinced that he had a severe auditory comprehension deficit.

One way to test whether a speech perception deficit could be the 'cause' of auditory comprehension deficits in Wernicke's aphasia is to see whether the degree of a speech perception deficit *predicts* how well a person understands language. To accomplish this, it is necessary to develop a speech perception test and an auditory comprehension battery. Testing a group of persons diagnosed with Wernicke's aphasia should reveal that as the speech perception deficit increases (as shown by increasingly poorer performance on the speech perception test), the worse should be the performance on the auditory comprehension test. And this is exactly what the results show (Robson et al. 2012a). The greater the speech perception deficit, the worse is auditory comprehension.

These findings tell us that that speech perception deficits can negatively affect auditory comprehension. Yet, this may not be the whole story. Recall from Chap. 2

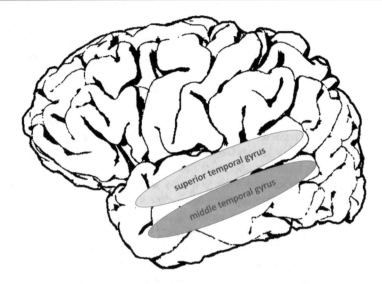

**Fig. 3.6** Lateral view of the left hemisphere showing the superior temporal gyrus involved in speech processing and the middle temporal gyrus involved in semantic processing

that one of the challenges in studying language and the brain in aphasia is that lesions resulting in aphasia tend to be large, and they are rarely restricted to one small narrow region. While those with Wernicke's aphasia have lesions in the superior temporal gyrus, they also often have lesions that extend to the middle temporal gyrus, a region that is adjacent to the superior temporal gyrus (see Fig. 3.6). The middle temporal gyrus has been implicated in neuroimaging research as being one of a number of areas involved in semantic processing (Binder et al. 2009; see also Chap. 8). Thus, it is possible that auditory comprehension deficits in Wernicke's aphasia stem not only from speech perception impairments but also from impairments in accessing the meanings of words (Robson et al. 2012b).

## 3.4   What Does it All Mean?

As we described at the beginning of this chapter, the sounds of language provide our connection to the outside world. This is clearly disrupted in aphasia – both in speaking and in understanding. There are obvious differences between speaking and understanding. Speaking requires motor commands to the articulators and perceiving requires acoustic analysis of the auditory input. Given the very different requirements and 'goals' for speaking and perceiving, it might be expected that there would be two separate and distinct systems and distinct patterns of impairment.

The facts of aphasia do not support this view. Deficits emerge in using the sounds of language for speech output and input in both Broca's and Wernicke's aphasia and as a consequence of lesions in distinct and separate neural areas (frontal and temporal lobes). The patterns of impairment across the two domains (speech output and speech input) are the same; there are more one feature errors than errors of more than one feature. This similarity suggests that a common neurally distributed network architecture underlies the sounds of language based on the organization of features in terms of their articulatory and acoustic similarity.

This connection between production and perception is essential to the normal use of language. We are both speakers and hearers. When you speak to someone, you not only produce speech but you are also perceiving it via feedback to what you are saying. You typically recognize within milliseconds when you have made a speech error. And it turns out you tailor the acoustic details of your production of speech to the characteristics of what you hear as others speak. Speakers adapt to what they hear and there is a convergence–a 'meeting of the minds' of sorts. For example, certain aspects of speech such as speaking rate 'more closely match' that of someone you may be conversing with (Cohen Priva et al. 2017).

The connection between what we say and what we hear is also essential in the development of language. Here, articulation and perception meet. Starting from infancy, children need an acoustic model to know that what they are articulating matches what they hear from the outside world. Their perception shapes their production. The importance of this connection is seen most clearly in those who are congenitally deaf. In this case, without an auditory model, speech production is often distorted and rarely reaches the clarity of someone with normal hearing.

While it may not seem obvious, we rely on this connection between perception and production in our everyday lives. Consider what happens when someone is asked to take their wedding vows or when someone is being sworn in to a position like President of the United States. They have to *repeat* what they hear. And we repeat all of the time – whether in response to learning that 'the car has a big dent in it', you might say 'the car has a big dent in it?', or asking a neuroscientist how to pronounce 'fasciculus', a part of the brain you had never heard of.

The connection between hearing and speaking is also manifest in aphasia. Both persons with Wernicke's and Broca's show some impairments in repetition, with particular difficulties for those with Wernicke's aphasia. There is one clinical syndrome, conduction aphasia, which has as its major presenting symptom a repetition deficit. Persons diagnosed with this type of aphasia generally show good auditory comprehension, so the speech information has clearly been analyzed and processed. Their speech production is fluent, but there are frequent errors in speech production, particularly in the occurrence of substitution errors. In fact, the pattern of errors in both speech perception and speech production mirrors what we have seen in Broca's and Wernicke's aphasia.

These findings once again suggest that the neural system underlying speech processing for both speech production and speech perception is broadly distributed. However, as we pointed out in Chap. 2 and in this chapter, because both anterior and posterior neural structures are recruited in processing speech and the pattern of impairment is the same does not mean that there are no functional differences in these areas. As we have seen, there are more speech production errors in Broca's aphasia and more speech perception errors in Wernicke's aphasia. And there are articulatory impairments in Broca's aphasia and none in Wernicke's aphasia. These differences reflect neural specialization in frontal and temporal areas of the brain. Frontal areas support production – the mapping from features and the sounds they represent into motor programs for articulation. Temporal areas support perception – the mapping of the fine acoustic detail to features and sounds. Thus, it is not surprising that despite similar patterns of breakdown in both speech production and speech perception, there are also differences that reflect those neural areas that serve as the gateways for the speech system to communicate with the external world through speaking and listening.

## References

Binder, J. R., Desai, R. H., Graves, W. W., & Conant, L. L. (2009). Where is the semantic system? A critical review and meta-analysis of 120 functional neuroimaging studies. Cerebral Cortex, 19(12), 2767–2796.

Cohen Priva, U., Edelist, L., & Gleason, E. (2017). Converging to the baseline: Corpus evidence for convergence in speech rate to interlocutor's baseline. The Journal of the Acoustical Society of America, 141(5), 2989–2996.

Fodor, J. A. (1983). Modularity of Mind: An Essay on Faculty Psychology. Cambridge, Massachusetts: MIT Press. ISBN 0-262-56025-9

Fromkin, V.A., 1971. The non-anomalous nature of anomalous utterances. Language, 47 (1), pp. 27–52. Stable URL: http://links.jstor.org/sici?sici=00978507%28197103%2947%3A1%3C27%3ATNNOAU%3E2.0.CO%3B2-M

Robson, H., Keidel, J. L., Ralph, M. A. L., & Sage, K. (2012a). Revealing and quantifying the impaired phonological analysis underpinning impaired comprehension in Wernicke's aphasia. Neuropsychologia, 50(2), 276–288.

Robson, H., Sage, K., & Ralph, M. A. L. (2012b). Wernicke's aphasia reflects a combination of acoustic-phonological and semantic control deficits: a case-series comparison of Wernicke's aphasia, semantic dementia and semantic aphasia. Neuropsychologia, 50(2), 266–275.

## Readings of Interest

Cutler, A. (2012). Native listening: Language experience and the recognition of spoken words. MIT Press.

Olmstead, A. J., Viswanathan, N., Cowan, T., & Yang, K. (2021). Phonetic adaptation in interlocutors with mismatched language backgrounds: A case for a phonetic synergy account. Journal of Phonetics, 87, 101054.

# Words, Words, and More Words: The Mental Lexicon

**4**

The mental lexicon is that dictionary in our head. We need it and use it constantly as we communicate. When we speak, we have to select the words we want in order to convey the meaning we intend. When we listen, we have to be able to match what we hear to a word in our dictionary. Only then, can we fully understand what is being communicated. Well, you say, what's the big deal?

It is a big deal if you consider that the estimates are that we know some 27,000–52,000 words (Brysbaert et al. 2016), and we have to get to them on the fly. We speak an amazing 100-150 words per minute which is about 2.5 words per second (https://virtualspeech.com/blog/average-speaking-rate-words-per-minute). This means we have to know what we want to say, select the words we want from the mental lexicon, put them together, and say them, all at warp speed. And listening to someone means we have to be able to analyze the sounds of speech and match the sequence of sounds onto words and their meanings in a matter of milliseconds.

How do we do this? It would be one thing were our dictionary to have some 40 or 50 words – but to be able to select words from thousands and process them in milliseconds requires not only a lot of brain power but also a system that allows us to access the words we want easily and efficiently. What if words in our mind and brain were organized like a dictionary? In this case, they would be in alphabetical order – so if you look up, for example, the word 'spool' in the dictionary (in this case, The American College Dictionary, Random House 1958, p. 1168), the words closest to it would be the one preceding it, 'spooky', and the one following it 'spoom'. The word 'spoom' meaning 'to run, or scud, as a ship before the wind' is definitely not in my lexicon. For me, the word following 'spool' that is in my lexicon is 'spoon'. Do you think that such an alphabetical list is the best and most efficient system or is there another way?

This chapter will try to answer this question as we examine through the lens of aphasia how the brain represents and organizes the words in our mental dictionary. Before starting, we need to think about what's in a word. That is, what information is part of a word entry?

© Springer Nature Switzerland AG 2022
S. E. Blumstein, *When Words Betray Us*,
https://doi.org/10.1007/978-3-030-95848-0_4

A word is essentially the nexus between a set of speech sounds, its meaning, and its grammatical use in a sentence. For example, the sounds associated with 'cat', 'c', 'a', 't', are connected to a meaning representation, in this case, an animal which is a feline with pointy ears, whiskers, and a tail. How the word 'cat' is used in a sentence is another crucial piece of information; 'cat' is a noun, 'the cat plays', and not another part of speech such as a verb, *'I cat every day' (the * indicates that the sentence is ungrammatical). We need all of this information in our mental dictionary to successfully communicate. Figure 4.1 shows how speech sounds, meaning, and syntactic role come together in the mental lexicon.

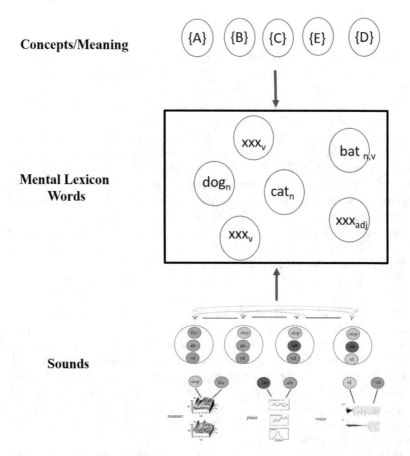

**Fig. 4.1** Schematic of the mental lexicon. Every word has a sound representation (bottom row) which connects to a conceptual/meaning representation (top row) as well as a syntactic role it plays in language (shown by subscripts on the lexical entry: $_n$ signifies noun, $_{adj}$ signifies adjective, and $_v$ signifies verb). Bottom left figure reprinted with permission from Mack, M. and Blumstein, S. E. 1983. Further evidence of acoustic invariance in speech production: The stop–glide contrast. The Journal of the Acoustical Society of America, 73(5), 1739–1750. Copyright 1983, Acoustic Society of America. Bottom middle figure reprinted with permission from Blumstein, S. E. and Stevens, K. N. 1979. Acoustic invariance in speech production: Evidence from measurements of the spectral characteristics of stop consonants. The Journal of the Acoustical Society of America, 66(4), 1001–1017. Copyright 1979, Acoustic Society of America

## 4.1    What's in a Name?: Naming Deficits in Aphasia

It is not unusual for either you or I to have difficulties in coming up with a word from time to time. It is frustrating to be sure, but the phenomenology that the word is right there, on the 'tip of our tongue', turns out to be quite accurate. As briefly mentioned in the introduction, despite our failure to come up with a particular word, we 'know' a lot about it. We often can identify its first sound, the number of syllables it has, what it means, and/or provide a synonym for it. For us, the word is not lost – it is there in our mental lexicon, just not immediately accessible.

It is no surprise then that word-finding would be a problem in aphasia. What is perhaps surprising is the extent of the problem. Indeed, words are particularly vulnerable in aphasia. Failing to come up with a word in a conversation or an inability to name the picture of an object or an action is common. Indeed, nearly all aphasics have word-finding impairments. Whether there is a difference between the severity and types of errors that occur is another issue and one which we will consider in some detail in the pages that follow.

Using words as we produce language requires that we get to them (access them) from our mental lexicon. In a conversation or even in a monologue, say a lecture by a professor, the speaker has a fair amount of latitude in the choice of words selected. The speech is spontaneous since it is created on the fly by the speaker. However, as I am sure you are aware, there are times when the speaker cannot immediately come up with the 'best' word for the context. What happens? One possibility is that the speaker stops the flow of speech and interjects that s/he cannot come up with the word. However, in most cases, speakers *circumlocute* – that is, they talk around the word, selecting other words and possibly phrases, to compensate for the temporarily 'lost' word. The only condition is that, in principle, the selected words make sense given the context.

There are situations, however, where the choice of words can be significantly narrowed. One of the easiest ways to constrain the possible choice of words is to ask someone to name a picture. For example, given a picture of a 'lamp', the response should be 'lamp', not 'furniture', 'bulb' or 'light', and it should certainly not be 'flashlight', 'thing', or 'shirt'.

Those with aphasia show word-finding problems in both spontaneous speech and in naming. Similar types of errors occur under both circumstances. However, as you might expect, it is easier to identify that an error has occurred and what that error is when asking someone to name a given object than when listening to a conversation.

Table 4.1 shows the variety of types of incorrect responses that someone with aphasia may produce in a naming task. While there seems to be a large variety of error types, there is actually a certain consistency to them. With the exception of the last two error types (unrelated word and neologism), to which we will return, the target word and error share a relationship in terms of their sound similarity, meaning similarity, or both. The crucial question is can we explain these errors? Where do they come from? How do they happen? Why do they happen?

**Table 4.1** Types of Naming Errors made in Aphasia

| Error type | Target (Picture) | Error |
|---|---|---|
| **Sound error** | Pear | Bear |
| **Meaning error** | Banana | Apple |
| **Sound + meaning error** | Prayer | Priest |
| **Semantic (meaning) blend of two words** | Shoe | Sheaker<br>Shoe + sneaker |
| **Sound error that is an unrelated word** | String | Stray |
| **Circumlocution** | Hamburger | Like to eat it |
| **Unrelated word** | Cup | House |
| **Neologism** | Ring | Sal |

## 4.2   Network Architecture of the Lexicon

Recall in Chap. 2 that we said that the components of language – speech sounds, words, and meanings – are each organized in a network-like architecture. Fortunately, the network architecture we talked about comes to our rescue/aid in accounting for the types of naming errors made in aphasia. When 'looking for a word' in our mental dictionary, we don't have to search the entire dictionary.

A simplified lexical network was shown in Fig. 2.6 in Chap. 2. As shown there, words are connected in the network based on their sound similarity, 'cat' and 'bat', and their meaning relationships, 'cat' and 'dog' or 'cat' and 'purr'. The greater the similarity or the closer the meaning relationship, the stronger the connection strength between words in the network. There are stronger connections between words that sound similar, 'cat' and 'bat', and words that share meaning attributes or are associated with each other, 'cat' and 'dog', than words that are only indirectly related or not related at all, e.g. 'dog' and 'purr' or 'mouse' and 'spoon'. Recall also that the activation of one word partially activates other words in the lexicon, and the connection strength between words reflects how closely their activation levels are in the network. The stronger the connection strength, the more likely one of the words will activate the other. For example, the picture of a 'cat' will activate both the sound and meaning representation of 'cat'. It will also partially activate words that sound similar to 'cat' like 'bat' and words that share a meaning relationship with 'cat' like 'dog' or 'purr'. Words that have a more distant relationship with 'cat' both in sound and meaning will be more weakly activated.

## 4.3   Where Naming Errors Come from

The network architecture comes to our aid when we are trying to retrieve a word either for speaking or for understanding. When 'looking for a word' in our mental dictionary, we don't have to search the entire dictionary. The network provides us with a limited set of possible options. This network architecture can also 'explain' the naming errors in aphasia shown in Table 4.1. Assume, as we did, when characterizing the effects of brain injury on language in aphasia in Chaps. 2 and 3, that the architecture of the network (its organization) is spared but brain injury introduces noise into the system. Noise in the lexicon will reduce the overall activation level of words, and it will also reduce differences in the activation levels between words. The result is that sometimes the word will be named correctly; other times it will not. When naming errors do occur, the particular error and error type may vary for the same word target. However, importantly, the errors will not be random. The closer two words are in the network either in terms of their sound and/or meaning properties, the greater the likelihood that the wrong word will be activated and ultimately selected over the target word, leading to a naming error.

Looking at the types of errors listed in Table 4.1, naming errors occur between words that share sound features but not meaning. For example, 'pear' is named for 'bear' and 'string' for 'stray'. Errors occur between words that share meaning attributes but not sound; 'banana' is misnamed as 'apple'. And errors occur between words that share both sound and meaning attributes; 'priest' is named for 'prayer'. There are even sound blends between two words that share meaning. An example, is 'sheaker' which is a response to the picture of a 'shoe'. This combination of the target word 'shoe' and the closely related word 'sneaker' shows that both words are partially activated, and individual sounds from each word are blended in the response.

What about the last three types of naming errors listed in Table 4.1? At first blush, these errors look like they challenge the claim that the structure of the lexicon and the lexical network are spared. But do they? What about the circumlocution 'like to eat it'. The word 'it' is empty of semantic content and could refer to anything that is inanimate, although being the object of 'eat' it narrows the possibilities to edible things. For most folks, 'like to eat it' is in the world of 'hamburgers' and indicates that the speaker is in the 'right semantic ballpark' but is unable to come up with the name of the object.

The production of unrelated words and neologisms when naming a picture is even more challenging. After all, the substitution of one word for another word that is neither similar in sound nor meaning with a word to be named looks like a random selection of one word for another. And the production of a neologism which is a nonword (a word that does not occur in English) appears to be 'a possible' but non-existent word; namely, the production of a possible sound sequence absent meaning. Again, this seems like a random error, one not a part of the lexical network.

Let's think further about the effects of noise on the lexical network as a possible hint about what may be going on with these two types of errors. The amount of noise introduced in the network reflects the extent of damage to the network much as the location, size, and extent of brain injury affects the severity of the language impairment in aphasia. Increased noise effectively reduces the distance between words within the lexical network, essentially 'spreading' the relationships between words more broadly. Words which are less closely related in sound or in meaning or not even related may now be mis-selected. A naming error of 'house' for the picture of a 'cup' reflects access to the lexicon under extremely noisy conditions. Another example of such a potential error is shown in Fig. 2.6 where 'spoon' is weakly connected to all of the other words in the lexical network by dint of it being a word in English.

Well, if being a word is required, what about neologisms? It turns out that although they are not words in the lexicon, they *could* be words. All languages have sound rules about what is an allowable or possible word. For example, 'string', 'sting', and 'sing' are all words in English. But *stling is not a possible word since the occurrence of 'stl' violates the allowable sound rules for English words.

We introduce new words into our language all of the time. As new discoveries or new products come into our culture, we add new words to refer to them. Think about all of the computer and internet terms we now use regularly – 'modem', 'malware', 'reboot'. These words were not a part of the daily vocabulary of the average person until the rise of the computer age. And imagine if you had invented a new product and wanted to name it. You have lots of choices – how about 'krim' or 'sim' or 'nove', or 'mestolie'. Any of those could be used so long as they follow the sound rules of the language – and if you are lucky and your product takes off commercially as did 'kleenex' or 'xerox', it would become a universal part of the mental lexicon of English speakers and beyond! Thus, neologisms produced in aphasia could be words but they just aren't.

So where do they come from? The extent of noise in the lexicon provides the answer. A large amount of noise can effectively diminish or break the tie between sounds and their meaning (review Fig. 4.1) leaving the sound component functioning independently of the meaning component. Neologisms are then not random productions. Rather, they reflect the generation of sound sequences that are possible words in English.

### 4.3.1 Naming in Aphasia

Now that we have suggested a basis for naming errors in aphasia, it is time to examine the extent to which actual naming behavior in aphasia lines up with theoretical predictions. As we indicated in the early part of this chapter, the types of naming errors shown in Table 4.1 do occur in aphasia. However, it turns out that there are differences between those with Broca's and Wernicke's aphasia in not only the frequency of naming errors but also in the likelihood that particular types of errors will occur.

In general, for both groups, naming errors occur which involve either a sound change, a meaning change, or both sound and meaning changes. Meaning changes are particularly evident when asking aphasics to name items grouped by categories. For example, in one part of the Boston Diagnostic Aphasia Exam (Goodglass and Kaplan 1972), a clinical exam used to assess language impairments in aphasia, drawings are presented to the patient grouped into different categories: objects (such as 'glove', 'key', 'chair'), letters (such as 'S', 'R', 'T'), and geometric figures (such as 'circle', 'triangle', 'square'). The examiner points to one of the items for naming. Typically, a subject who makes a naming error on this task will name the wrong object, but it will be drawn from the correct category. For example, instead of saying 'circle' when the examiner points to its picture, the response might be 'triangle'. It would typically not be 'S'.

These types of errors highlight the network architecture of the lexicon and the effects of noise on accessing individual words. The words in the lexicon are still 'there'. Noise in the network results in errors that reflect the similarity of the words, in this case words which belong to the same category.

Indeed, case studies looking at individuals who typically do not present with aphasia have shown that brain injury can result in category-specific deficits. In particular, selective impairments have been found in naming animals, in naming fruits and vegetables, and in naming tools (Berndt 1988). In the worst case, degenerative neurological disorders such as dementia show a gradual loss of semantic knowledge (Saffran and Schwartz 1994).

Such category-specific impairments and loss of characteristics of particular categories of words raise interesting and important questions about the neural basis of categories and the concepts or meanings that underlie them. Does the brain organize objects in terms of categories? That is, is there a category for tools, for musical instruments, for computer parts, or does the brain organize objects in terms of your experience with them – how you perceive them, how you interact with them, and/or how you are used – reflecting their perceptual, motor, and functional properties? For example, the perceptual attributes of a piano include how it *looks* by eye, how it *sounds* by ear; its motor attributes– what it feels like when you *touch* the instrument or *play* it; and its functional properties – making music. Presumably, all of the above factors play a role in defining what the meaning of a word is. If we were to assume, as we have proposed, that words and their meanings are organized in terms of a network of shared features or properties, then the system would operate *as if* it were organized by categories, but in actuality it is not. Rather, it is the architecture of the network that gives rise to 'category-like' behavior. Categories *emerge* from the structure of the network, but they are not defined by nor are they an inherent part of the network.

Lets' get back to naming deficits in aphasia. As we have said, both Broca's and Wernicke's aphasics have naming deficits and make both sound and meaning errors. However, the nature of the naming deficit is not the same in the two syndromes. Those with Wernicke's aphasia display a more severe naming impairment, making considerably more errors than those with Broca's aphasia. The more severe their deficit, the more likely the presence of unrelated word errors and neologisms. In contrast, these two error types rarely occur in Broca's aphasia.

Why would severity of naming impairments be more acute in Wernicke's aphasia than Broca's aphasia? Recall that the severity of a deficit in aphasia is a reflection of the amount of noise in the network. Given that the potential source of unrelated word substitution errors and neologisms is due to increased noise, we would expect that these types of errors would occur in the context of a severe naming deficit. And that is exactly what typically happens for those who have a severe Wernicke's aphasia. Because of the interconnectedness of sound and meaning in the lexicon, a large amount of noise in the network will result in the partial activation of more distant words which may be unrelated in meaning and sound from the word to be named. As for neologisms, at the extreme, neologisms are productions in which there is so much noise that the connection between sounds and their meanings is weakened leaving the activation of sounds that do not correspond to any actual words in the lexicon.

There is another possible source of neologisms. They could originate from the selection of a wrong word produced with multiple sound errors. Let's say a Wernicke's aphasic is asked to name the picture of a 'fork' and says instead 'mas'. How could this be? In this example, the semantically related, but incorrect word 'knife' may be activated along with a series of sound substitutions; 'm' substitutes for the 'n' of 'knife', 'a' substitutes for 'i', and 's' substitutes for 'f' resulting in the neologistic response 'mas'.

Neologisms are interesting because while they are possible words, they do not become usable words for those with Wernicke's aphasia. In other words, someone with Wernicke's aphasia who produces a neologism will not produce it again nor will it be recognized if said by the examiner. Moreover, the speech output of those with Wernicke's aphasia is not all neologisms. Rather their speech is peppered with them. Most of the output includes content words like 'book' and 'run', grammatical words like 'the', and grammatical endings attached to word stems like the past tense marker 'ed' on 'walked'. For example, a typical production might be 'I played over there to the pakoli'.

### 4.3.2  It's Right There on the Tip of My Tongue: And It Is!

When you or I are unable to come up with a name, we are often able to succeed in producing the word if given a cue such as a word that is similar in meaning, a semantically related context, or the first sound or sounds of a word. That we were able to retrieve the word tells us that the elusive word is simply inaccessible for the moment.

Does the same hold true in aphasia, where naming errors or simply a failure to come up with any response may be common? Does a failure to name during an aphasia exam or in conversation mean that the word is no longer an active part of the aphasic's vocabulary? Or does it mean that there is a deficit in accessing the meaning of a word or alternatively accessing the sounds of a word? The variability in performance in naming suggests that the word is really still 'there' in the lexicon. As mentioned earlier, a naming error on a word or a failure to name it at all at one time may be produced correctly at another time. But what about the moment when the examiner presents a picture for naming and there is a failure to name it correctly? Is that elusive word still there, just not easily accessible because of noise in the system? It turns out that just as for us, presenting a 'hint' to either the meaning or sound of a word often helps trigger its name for someone with a naming deficit in aphasia. Say you show the picture of a 'bear' to someone with aphasia, and s/he cannot come up with the name. There are several 'hints' you can give. One is a semantically relevant context like 'Brown University's mascot is a brown ___' or 'Smokey the ___'. Another is the presentation of the first sound of the word; 'b...' or the first consonant and vowel 'bæ...'. Either type of hint can result in a successful naming response. Such success tells us that, despite a failure to name a word at a particular point in time, both its meaning and sound attributes are 'alive'. As you might expect, however, there are differences among those with aphasia in the extent to which such 'hints' help in accessing a word to be named. The greater the severity of the aphasia and the greater the auditory comprehension impairment, the less likely will be the ability to take advantage of either meaning or sound cues to come up with the correct name.

Up to now we have talked about how someone with aphasia can access a word to retrieve its name. But the lexicon is just as critical for recognizing spoken words when you hear them. Imagine if you could not recognize words or you recognized them incorrectly. This would obviously have a devastating effect on your ability to understand language. Let us now turn to how brain injury in aphasia affects the lexical network and the connection between sounds and their meanings in word recognition. Based on what you have learned about the architecture of the lexical network (review Fig. 2.6), can you make any predictions about how word recognition will be affected in aphasia? Let's see if you are correct.

## 4.4   Recognizing Words

As we have shown in our discussion of naming deficits in aphasia, words are connected in a lexical network based on their sound similarity and meaning relationships. Given a single lexical network which is used for speaking and for understanding (as shown in Fig. 2.6), we would expect similar patterns as we saw for naming to emerge in word recognition as a result of brain injury in aphasia. In particular, because of the network properties of the lexicon, we would expect that sound and meaning similarity would be particularly vulnerable in recognizing words.

We know this to be the case when we present words auditorily and ask those with aphasia to select the presented word from a set of words that may be similar in either sound or meaning. Errors can occur, and when they do, aphasics will point to a word that sounds similar, for example, hearing 'pear', pointing to 'bear', or they will point to a word that is related in meaning, for example, hearing 'pear', pointing to 'apple'.

This tells us about the pattern of errors – but not whether in recognizing words, the organization of the lexicon in aphasia is similar to those without brain injury. This is particularly important when considering the severe auditory comprehension deficit in Wernicke's aphasia we discussed in Sect. 3.3.2 in Chap. 3. Is it possible that the deficit is so great that the underlying meaning/concepts of words are impaired or lost?

In order to answer this question, we can first turn to how researchers have studied word recognition in folks without brain injury. One way is to present subjects with a stimulus and ask them to make a decision about whether the stimulus is a word or not by pressing one of two response buttons, one button if the stimulus is a word and the other button if the stimulus is a nonword. Note, in completing this task, the subjects do not have to say what the stimulus is, what it sounds like, or what it means. They only have to recognize that it is a word in the lexicon or not. When subjects are asked to press a button as they make such a *lexical decision*, we can get two measures – one measure is whether the subject's response is correct, and the other measure is how quickly the subject responds.

What if instead of presenting only one stimulus, we now precede the lexical decision *target* with another stimulus, called the *prime*. Table 4.2 shows the stimulus conditions that can be used in a possible experiment. Here, the word target can be preceded by a word that is semantically related to it or the word target can be preceded by a word that is semantically unrelated to it. Note that in this example, there is a third nonword condition. Here, the target stimulus is a nonword preceded by a word. Why is this condition necessary? Imagine if the target stimulus were always a word. Would subjects really have to make a decision about whether the stimulus was a word or not? No. They would learn that all target stimuli are words, and they would simply press the word button all of the time without needing to listen to and process the stimuli.

Now let's think again about the architecture of the lexical network and consider how the network will 'respond' when the prime-target pair is semantically related compared to when it is not. Figure 4.2 illustrates this scenario. Presentation of the prime word 'fruit', activates 'fruit' and partially activates semantically related words like 'apple', 'pear', and 'cherry'. In contrast, 'fruit' only weakly activates 'shoe' since it only shares with 'shoe' the fact that it is a word in the lexicon.

There are behavioral consequences of such an architecture. Given prime-target pairs for lexical decision, it takes less time for the subject to decide that 'apple' is a word when it is preceded by the semantically related word 'fruit' than when it is preceded by the semantically unrelated word 'shoe'. This is because "apple' has been partially activated by "fruit' even before it is presented to the subject for lexical decision. In contrast, because there is no semantic relationship between 'shoe' and 'apple', the determination of whether 'apple' is a word or not is based solely on the subject's accessing 'apple' from the lexicon in order to make the lexical decision.

*Semantic priming,* faster reaction-time latencies for semantically related words compared to semantically unrelated words, as this effect is called, is robust and easily shown in individuals without brain injury. Semantic priming also occurs in both Broca's and Wernicke's aphasia. These findings are particularly critical when considering those with Wernicke's aphasia who have severe auditory language comprehension deficits. That they show semantic priming indicates that despite their injury, the architecture of their lexicon is intact. Because of the severity of their disorder, those with Wernicke's aphasia do make more errors in deciding whether a target stimulus is a word or not. However, when they recognize a word in the lexicon, the organization of the lexical network is similar to those without brain injury.

**Table 4.2** Examples of stimulus conditions and stimuli used in a semantic priming experiment

|            | Related | Unrelated | Nonword |
|------------|---------|-----------|---------|
| **Prime**  | Fruit   | Shoe      | Nose    |
| **Target** | Apple   | Apple     | Plub    |

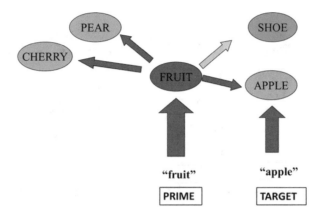

**Fig. 4.2** Response of the lexical network in a semantic priming experiment. The prime stimulus 'fruit' activates the lexical network. The degree of activation of the words in the network varies as a function of their semantic relationship to the prime stimulus 'fruit'. There is the greatest activation for 'fruit' in the network, less activation for words semantically related to 'fruit' ('cherry', pear' and 'apple'), and minimal activation for 'shoe' which is semantically unrelated to 'fruit'. When 'apple' is presented as the target stimulus, it is already partially activated in the lexical network by 'fruit'. See text for behavioral consequences

### 4.4.1 The Eyes Have It

As we have discussed, words are connected in the network not only in terms of their meanings but also in terms of their sounds. Again, the network organization of the lexicon suggests that when the sounds of a word are activated, they partially activate other words that sound similar to it. Just as aphasia does not affect the network of meanings during word recognition, it does not affect the network of connections based on their sound similarity. This can be shown by tracking subjects' eye movements when they are asked to look at or point to an object presented along with other objects. The beauty of this method is that it is not only possible to track how the subject recognizes a word over time, but it provides a window into how the subject scans the objects in the array.

Figure 4.3 shows an example of a stimulus array presented to the subject. Let's say the subject is asked to point to the 'lamb'. Note that 'lamb', located in the top left corner of the array shares all but the last sound with 'lamp', located in the bottom right. The other two drawings, 'anchor' and 'truck' are not related to 'lamb' either in sound or meaning. So what do you think happens?

When subjects without brain injury are presented with such a stimulus array they look consistently at the target picture, in this case 'lamb', within a few hundred milliseconds. However, as they are scanning their choices and before they consistently look at 'lamb', they look longer at 'lamp' than at the other two unrelated pictures, in this case 'truck' and 'anchor'.

This pattern of eye movements reflects the network architecture of the lexicon shown in Fig. 4.1. Hearing the target word 'lamb' activates its sounds and also

**Fig. 4.3** Stimuli used in an eyetracking experiment. 'Lamb' and 'lamp' share all but the last sound of the word; 'anchor' and 'truck' share neither sound nor meaning with each other nor any of the other stimuli

partially activates the sound representation for similar sounding words like 'lamp'. Both words are activated, and, in this sense, they are 'competing' with each other. It is this competition with the target stimulus which results in more looks to 'lamp' than to 'truck' and 'anchor', words which do not compete with 'lamb' in either sound or meaning.

With respect to aphasia, both those with Broca's and Wernicke's aphasia successfully recognize the target word, although not as accurately as those without brain injury. Critically, they show competitor effects with more looks to the similar sounding competitor than to words unrelated in sound and meaning. The presence of such competitor effects supports the integrity of the architecture of the lexical network in both those with Broca's and Wernicke's aphasia.

However, all is not normal. Remember, brain injury introduces noise in the network and its effects are to reduce the activation level of words and the differences between them. The result is that noise renders the performance of the network 'sluggish' in selecting a target word from those words that are similar to it. This is exactly what happens in aphasia. It takes longer for aphasics to look at the target, and during this interval, they look more at the competitor word than at the two unrelated words. It is as though the word recognition system in aphasia is having 'difficulty' in resolving the competition between words in the network. As one might expect, those with Wernicke's aphasia are slower in recognizing the target word, and they look even longer at the word competitor than those with Broca's aphasia, again reflecting the greater severity of their lexical deficit.

## 4.5    One More Word

There is one important take-away from this chapter about lexical deficits in aphasia. It is the following: there is a common lexical network that supports accessing words for speaking and for understanding. As we have described, words are organized in a network-like architecture connected with each other in terms of their similarity in both meaning relationships and sound similarity. Noise in the network resulting from brain injury in aphasia affects its performance. Closely related words (in meaning and sound) are more likely to be incorrectly accessed in both naming and word recognition. This single lexical network is analogous to what we saw in Chap. 3 where a common speech network serviced both speaking and understanding.

The notion that a single lexicon underlies speaking and understanding is not without its critics. There are some who argue that there are two separate lexicons – an output lexicon accessed for speaking and an input lexicon accessed for understanding (Caramazza 1988). How does one decide what is the best characterization of the lexicon? Are there one or two lexicons?

When there are two theories or explanations, *Occam's Razor* may be utilized to choose between them. Occam's razor is a principle that originated from philosophy and is applied throughout the sciences. Assuming that more than one solution can account for the data, it states that given two solutions or explanations to a problem or theory, the one that makes the fewest assumptions and, in this sense, is the simplest is the preferred one.

So let's consider the implications for having two lexicons – an input lexicon and an output lexicon. Having two separate lexicons doubles the size of the lexicon, and thus requires more storage and more memory. It also requires two separate architectures – one for input and one for output. An additional mechanism is needed to connect the words in the two lexicons. This is because when you speak a word, you hear it as you produce it – the spoken and heard word 'match'. And if you are talking about the same thing with someone else, what you say and what you hear them say have to map on to the same representation. For example, as a speaker-hearer of English, the word 'dog' in the input lexicon has to have an equivalent entry for the word 'dog' in the output lexicon.

Without question, two lexicons are more complex and include more assumptions than having a single lexicon. At this point, the evidence points to a single lexicon. However, science always seeks more evidence and more data, and you should, as will I, keep our eyes open for challenges to the idea that there is a single lexicon.

## References

The American College Dictionary (1958). New York: Random House.

Berndt, R. S. (1988). Category-specific deficits in aphasia. Aphasiology, 2(3–4), 237–240.

Brysbaert, M., Stevens, M., Mandera, P., and Keuleers, E. (2016). How many words do we know? Practical estimates of vocabulary size dependent on word definition, the degree of language input and the participant's age. Frontiers in Psychology, 7, 1116.

Caramazza, A. (1988). Some aspects of language processing revealed through the analysis of acquired aphasia: The lexical system. Annual Review of Neuroscience, 11, 395–421.

Goodglass, H., and Kaplan, E. (1972). The assessment of aphasia and related disorders. Lea & Febiger.

Saffran, E. M., and Schwartz, M. F. (1994). Of cabbages and things: Semantic memory from a neuropsychological perspective—A tutorial review. In C. Umiltà & M. Moscovitch (Eds.), Attention and performance 15: Conscious and nonconscious information processing (pp. 507–536). Cambridge: The MIT Press.

## Readings of Interest

Dell G.S., Schwartz M.F., Martin N., Saffran, E.M., and Gagnon, D.A. (1997). Lexical access in aphasic and nonaphasic speakers. Psychological Review, 104(4), 801–38. doi: https://doi.org/1 0.1037/0033-295x.104.4.801.

Mirman, D., Strauss, T. J., Brecher, A., Walker, G. M., Sobel, P., Dell, G. S., et al. (2010). A large, searchable, web-based database of aphasic performance on picture naming and other tests of cognitive function. Cognitive Neuropsychology, 27,495–504. doi: https://doi.org/10.108 0/02643294.2011.574112

# Putting Words Together: Syntax

**5**

In the previous chapter we talked about words and the effects of brain injury on producing and understanding them. Is that all there is to language – words, words, and more words (and of course, as we discussed, their meanings). In some sense, yes. We typically string words together in sentences to convey a message, and one word is rarely sufficient for communication.

Indeed, it is our ability to produce and understand new and novel sentences that allows language to be infinitely creative. Think about it. If we could stop time and think about language at this very moment, we would have a finite inventory of sounds of a language as well as a finite inventory of words. However, we create new words all of the time and introduce them into our language to reflect new concepts and additions to our culture and to our world. Take the word 'google', for example. It was first used in 1997 (https://www.etymonline.com/word/google) and is now a staple of our (or at least my) vocabulary – it can be used as a verb in all of its tenses, 'I *google* all of the time, but I know I *googled* him yesterday'; as a gerund, '*Googling* is fun'; or as a proper noun, 'I looked it up on *Google*'.

In contrast to sounds and words, there are a potentially infinite number of different sentences that we can produce or that we hear at any time. I bet that this sentence that I am typing on my laptop this very moment while sitting with my dog in the family room has never been produced before and may not be produced ever again. So much for the creativity of language!

Why are sentences so important? After all, they are simply words strung together. Let's do a *gedanken* experiment – a thought experiment in which we just try to think through a problem and try to solve it with no experimental apparatus, stimuli, or subjects. The question to be answered is whether language communication could 'work' were we to depend solely on words produced in isolation. Here we go: 'school'. What information was I trying to convey? Was I trying to tell you 'I am going to school today'; 'I love school'; 'My school is 10 miles from my house'; or something else? Absent context or some prior knowledge you may have, there is no way you can tell. It is obvious that language is much richer and requires that words be put together into sentences.

© Springer Nature Switzerland AG 2022
S. E. Blumstein, *When Words Betray Us*,
https://doi.org/10.1007/978-3-030-95848-0_5

## 5.1    Structure in the Strings

Making and understanding sentences seems like an easy task. It may seem as easy as simply putting words together in a linear string – one word following another. But is this true? Much research has shown that even though we produce or hear the words in sentences linearly, there is a structure to them. First and foremost, the order of words is not free – I cannot say 'book the on table the is' for 'the book is on the table'. Every language has syntactic 'rules' that constrain what the order of words in sentences can be. For example, to describe the color of a book in English, we would say, 'the red book'. In French, the adjective 'rouge' meaning 'red' follows the noun 'livre' meaning 'book', 'le livre rouge'. The same rule applies in Spanish, 'el libro rojo' translated literally as 'the book red'.

Word order is not the only constraint placed on sentences. Sentences are not simply made up of a string of words in a particular order. Words within a sentence have a structure to them; they group together into larger units called *phrases* based on the syntactic role they play. How phrases go together provides information about the meaning of the sentence. For example, Fig. 5.1 shows the syntactic organization of the sentence 'the boy kisses the girl'. Knowing the syntax of the sentence tells you who was the doer of the action of kissing ('the boy') and who was the receiver of the action 'the girl'.

You may say, all well and good. That is what linguists say about how syntax works. But for me, I do not do that. Well, in fact you do. Indeed, whether you

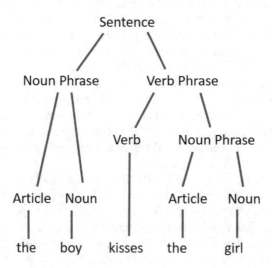

**Fig. 5.1**  Schematic of the grammatical organization of the sentence 'The boy kisses the girl'. A sentence is organized here as a phrase structure tree showing the hierarchical organization of the sentence. The sentence is made up of a Noun Phrase and a Verb Phrase. Each phrase in turn can be broken down into grammatical constituents or parts. The first Noun Phrase is made up of an Article (the) and a Noun (boy). The Verb Phrase is made up of a Verb (kisses) followed by the second Noun Phrase in the sentence which consists of an Article (the) followed by a noun (girl)

produce or listen to sentences, you break words up into groups or phrases, and you do this automatically without even thinking about it. All of us have an unconscious knowledge of the syntax of our language and we use it to both produce and understand the meaning of sentences. Let's look at several examples which demonstrate that you are indeed a linguist and have an implicit knowledge of syntax.

If I were to ask you to read aloud the sentence 'the cat chases the mouse' and to put pauses in the sentence as though you were talking to someone who seems to be having trouble following what you are saying, you would likely say, 'the cat… chases… the mouse'. You would not say, 'the…cat chases the mouse', or 'the cat chases … the mouse', or 'the cat chases the … mouse'. In other words, you would break the sentence down into its syntactic parts or phrases.

The syntax of what you hear when someone produces a sentence also influences what you say when you produce one yourself. If you were to hear a sentence like 'the waiter serves the woman a hamburger' and then you were asked to describe a picture which shares neither words nor meaning with the heard sentence like the one in Fig. 5.2, you would likely say 'the boy gives the girl a book'. You could have conveyed the same meaning with a sentence having a different syntax and said: 'the boy gives a book to the girl', but you didn't. Why?

Perhaps, it is simply because 'the boy gives the girl a book' is more preferred by you or is a more common usage. As logical as this may sound, that is not the explanation. This can be shown by the following. If rather than hearing the sentence 'the waiter serves the woman a hamburger', this time you hear the sentence 'the waiter serves a hamburger to the woman', and are again asked to describe Fig. 5.2, you

**Fig. 5.2** Example of a typical test stimulus used to elicit sentence responses from subjects

would now more likely say, 'the boy gives a book to the girl' rather than 'the boy gives the girl a book'. So you now have said two versions of the same sentence indicating that something other than your preference or sentence frequency is going on.

Ideas? What is common between the sentence you heard and the sentence you used to describe the picture is that they share the same syntactic structure. This means that you did more than simply listen to the words and meaning of the heard sentence. Rather, you automatically parsed the syntactic structure of the sentence – you broke it down into abstract syntactic phrases, and, unaware, you mirrored it when describing the picture (Bock 1986).

There is other evidence that even if unawares, we process the abstract structure of a sentence. Not only does syntactic structure affect how we plan and produce sentences as shown in the previous example, but it also affects how we process sentences as we hear them and remember them. We do better remembering words in sentences we hear when the words are in the same syntactic phrase than when they are in different phrases. For example, you are more likely to recall that 'Susan' follows the word 'help' in sentence b. than in sentence a.

(a) In her plan to help Susan was very optimistic.
(b) Her plan to help Susan was very optimistic.

Why? After all, in both sentences the two words 'help Susan' are next to each other, so if you were processing the words in these sentences in a linear order, there should be no difference in remembering Susan in the two sentences. But note that, although next to each other, in sentence a., 'help' and 'Susan' are part of two different phrases ('In her plan to help…Susan was very optimistic'). In contrast, in sentence b., the two words 'help Susan' are in the same syntactic phrase ('Her plan to help Susan…was very optimistic'). The results show that it is easier to remember words that are a part of the same phrase than words that are a part of different phrases (Garrett et al. 1966). Again, this means that as you are listening to sentences, you are automatically analyzing the syntactic structure of the sentence.

Knowledge, even implicit, of the grammatical or syntactic structure of a sentence is critical for language communication. It is your knowledge of syntax that lets you understand who the lucky person is in the sentence 'John gave Mary a million dollars'. The syntax of English tells you who is the giver of the action, who is the receiver of the action, and what was given. Thus, grammar is a crucial vehicle for conveying meaning.

In this chapter, we will consider how brain injury in aphasia affects the use of syntax in speaking and understanding. Just as words in the lexicon connect form, their sound structure, with function, their meaning, syntax connects form, its syntactic structure, with function, its meaning. Form and function are inseparable and, as we shall see, together they form a network of connections necessary for language production and comprehension.

## 5.2    Problems with Grammar in Aphasia: Sentence Production

Without question, brain injury in aphasia affects the use of grammar in both speaking and understanding. For both Broca's and Wernicke's aphasia, there is a reduction in the complexity of sentences produced and increased difficulty in comprehension of complex sentences. We will consider sentence complexity a little later in this chapter. Before doing so, we will focus on a feature of the sentence production of Broca's aphasia that has often been considered to be the hallmark of this syndrome. In addition to nonfluent speech output, those with Broca's aphasia tend to omit the little grammatical words in English such as 'the', 'is', be', 'have' as well as grammatical endings at the ends of words. For example, the present and past tense of a verb like 'walks' and 'walked' might be produced as 'walk' in both cases. And sentences like 'John is running' or 'Sue will be going to school' may be produced as 'John run' and 'Sue go school' respectively. As a consequence, the speech output in Broca's aphasia is often described as *telegraphic* or *agrammatic*.

Appellations or descriptions like 'telegraphic' or 'agrammatic' are often useful, as they describe the deficit. However, descriptors can also be misleading. They often imply that the term provides an explanation of the problem. Let's consider first the term telegraphic as a descriptor for the speech output deficit of Broca's aphasia. Think about what you would do if you wanted to send a telegram or a text to someone who did not know your travel plans telling them that your plane has been rerouted to a new destination: 'surprisingly, my plane will be arriving in Paris at 10 in the morning'. Seems simple enough, but there is a hitch; to send the telegram or text you have to pay by the letter. So to convey the information but to keep it as inexpensive as possible, you would pare down your sentence to the essentials. You might say something like 'ugh, plane arrive Paris 10 AM'. Sounds a little like someone with Broca's aphasia.

What if telegraphic speech is, as some have suggested, the result of a conscious strategy overtly used by those with Broca's aphasia to compensate for their difficulty in producing speech by saying only what is minimally necessary to convey meaning. To avoid paying the 'high price' of producing fully grammatical sentences, they select the semantically informative pieces of sentences such as nouns, verbs, and adjectives, and omit the grammatical endings of words and grammatical words such as 'the', 'and', and 'is'. This is a good working hypothesis as an explanation for why Broca's aphasics are telegraphic. However, the use of the term telegraphic speech may give a false impression of the nature of the deficit in Broca's aphasia suggesting that it reflects an effort to circumvent the difficulty of producing a lot of speech while still conveying their message. As we will see below, evidence suggests that although non-fluency and articulation deficits may contribute to the syntactic deficit of Broca's aphasia, at its core, telegrammatic speech is a syntactic deficit.

What about the term agrammatism? Is this a fair descriptor of the output in Broca's aphasia? Here, there are a number of caveats that suggest there is more to this story. First, agrammatism in Broca's aphasia does not mean 'no' or 'without' grammar (which is the literal meaning of agrammatism). Critically, although often omitted, grammatical words and endings are <u>not</u> lost. Just as we learned that neither sounds (in Chap. 3) nor words (in Chap. 4) are lost in aphasia, neither is it the case for grammatical words and grammatical endings in Broca's aphasia. For those with Broca's aphasia who can string a number of words together (even a few words), grammatical words may appear in their speech. For example, it is not uncommon to witness the use of the article 'the' in a phrase such as 'the baby', or grammatical endings such as the present tense ending '-s' in the sentence 'he wants'. Thus, although Broca's aphasia is characterized by clear difficulties in the use of the grammatical features of language, performance in their usage is variable not absent.

Even more telling is consideration of languages other than English. Why would this matter? A critical question to ask is what happens to speakers of languages other than English who present with Broca's aphasia. Do they have agrammatism, and what form does it take? There are thousands of languages in the world, and, as we describe below, they have very different ways of implementing grammar. This allows us to ask whether agrammatism and telegrammatism emerge in Broca's aphasia across languages.

## 5.2.1  Agrammatism around the World

There are many differences in the form that syntax takes across language. This underscores the importance of not focusing on one language in trying to understand its neural basis. Looking at aphasia in different languages provides a window into whether damage to the neurological structures that give rise to agrammatism in English also result in similar patterns of agrammatism in other languages. If differences emerge it will tell us that the 'hows' of a language's syntax has different neural substrates, and it is the form and not its functional role that guides its neural representation. If similar patterns emerge, it tells us that it is the functional role of syntax not its form that guides its neural representation.

Let's briefly review a few different languages, in particular, German, Japanese, Hebrew, and Arabic. These languages realize syntax in very different ways. You do not need to know the details. However, it is important to get a sense of the variety that exists across languages of the world before we consider what happens in aphasia.

English is basically a *word order language*. Grammatical differences are conveyed by the order of words in the sentence, e.g. 'the clown chases the midget' vs. 'the midget chases the clown'. In this example, the word order tells us who is doing the chasing and who is being chased. Note that there are several types of syntactic markers. Some like 'the' are independent words. In this sense, they are *free* and can stand alone. Other grammatical markers in English are added to a word to specify its syntactic role. These markers cannot stand alone and are *bound* to the stem of the word they are attached to. For example, '-s' is the marker for the plural in English and is added to the end of a noun such as 'book' to form 'books'. One cannot say 's' in isolation to indicate plural. Similarly, 'ed' is a past tense marker bound to a verb stem such as 'walk' to form 'walked'. Like the plural, the past tense marker cannot stand alone. Grammatical markers, whether free or bound, play a smaller role in English than in many other languages.

German is one such language. It is an *inflectional language* in which syntactic relations are conveyed by endings or inflections on words to signal the grammatical role of the word in the sentence. Inflections are bound to the words. For example, the stem of the verb 'go' is 'geh' and bound grammatical endings are added to it to indicate who is doing the going; I go: 'gehe', 'you go': gehst', he goes: 'geht'. Importantly, 'geh' cannot occur alone, nor can the grammatical endings indicating I: '-e', you: '-st', and he: 't'. The two parts of the word, the stem and inflectional ending are 'glued' together! Similarly, the word 'the' in German has bound grammatical markers attached to it indicating whether the noun it modifies is the subject, 'der', or the object, 'den', of a sentence. Because such syntactic information is a part of the word, word order is less strict in German. For example, for the sentence 'the man sees the dog', it is possible to say 'D*er* ('the' subject) Mann (man) sieht (sees) d*en* ('the' object) Hund ('dog'), 'D*en* Hund sieht d*er* Mann', or 'D*er* Mann d*en* Hund sieht'. In English, changing word order changes the meaning of the sentence – 'the man sees the dog' does not mean the same thing as 'the dog sees the man'.

In contrast to German and English, Japanese is an *agglutinative language* in which particles follow the words they modify. It is these particles which convey syntactic information. The sentence 'the man sees the dog' is realized in Japanese as 'otoko (man) wa (topic or subject) inu (dog) o (object) miru (sees)'. In this way, both words and particles can stand alone, although the particles cannot occur alone in a sentence without a word it modifies. Note too that word order in Japanese is not 'subject-verb-object' as in English, but it is typically 'subject-object- verb'.

Finally, Arabic and Hebrew are Semitic languages that convey grammatical information by infixing different vowels around the consonant root which corresponds to the word. For example, the root for the verb 'write' is 'ktb' in Arabic and 'ktv' in Hebrew. Vowels are added to mark whether the word is present tense, 'koteb' and 'kotev' in Arabic and Hebrew respectively or past tense 'katab' and 'katav'. The three consonants defining the verb cannot stand alone, nor can the vowel grammatical markers.

As you can see, there are many differences in the form that syntax takes across languages. This underscores the importance of not focusing on one language in trying to understand whether damage to the neurological structures that give rise to agrammatism in English also results in similar patterns of agrammatism in German, Japanese, Hebrew, and Arabic speakers. Let's consider some findings.

Agrammatic German speakers may omit articles just as in English. However, grammatical endings are not omitted or lost, but rather one marker is substituted for another. For example, the verb 'geht', 'he goes', might be produced as 'to go', 'gehen'. Furthermore, a German-speaking agrammatic might rely on subject-verb-object word order, rather than using the article which is marked for whether it is the subject, 'der', or object, 'den', of the sentence.

Perhaps more striking, Hebrew and Arabic speaking agrammatics do not 'lose' or omit grammatical particles – if this were the case, the three consonant root forms of words such as 'ktb' or 'ktv' would be produced in isolation. This is not allowed in these two languages. Rather, agrammatic speakers of Semitic languages use a semantically simpler grammatical form such as the present tense 'koteb' of the verb in Arabic instead of the past tense, 'katab'.

What does this tell us about English agrammatics? Is English different, omitting grammatical endings, and German and Arabic speakers, in contrast, substituting grammatical endings? Let's rethink the example of the English-speaking agrammatic aphasic who said 'walk' instead of 'walked'. Does the production of 'walk' mean there is no ending on the verb? In actuality, 'walk' is the present tense of the verb 'walk'. Grammatically, the ending for the present tense of the verb 'walk' in the sentence 'I walk' is the verb stem plus the present tense ending. In English, that ending is actually, zero, ø. That is, the absence of a grammatical marker on 'walk', in this case 'walkø'' indicates the present tense of the verb in English. Thus, when 'walk' is produced instead of 'walked' in agrammatic speakers, it is still an acceptable verb form in English. Similarly, the plural for the word 'books' might be realized as 'book'. What is the grammatical marker that corresponds to the singular form of most nouns in English? It is also ø; 'book ø' means 'a singular book'. Thus, the production of 'book' rather than 'books' results in an acceptable word form. What do these results tell us? It appears then that there is no difference between the patterns of agrammatism across languages. English agrammatics *substitute* grammatical endings as do German, Hebrew, and Arabic agrammatics.

Overall, results of studies of agrammatism across languages have shown remarkable consistency in the patterns of breakdown. Among them, there is a reduction in the syntactic complexity of sentences, and a tendency to follow a particular word order when producing sentences. Syntactic errors do not occur all of the time but when they do, there are omissions of free grammatical words and substitutions of bound grammatical markers. Critically, errors follow the 'rules' of the language. Just as we saw in Chap. 3 looking at speech production errors and Chap. 4 looking at word errors, substitution of syntactic markers results in possible word forms in the language.

Taken together, such findings suggest that there is a neurological substrate that is functionally specialized for syntax, and this specialization corresponds to the grammatical properties inherent in each language. As we will see in Chap. 7, the same neural substrate when damaged results in agrammatic speech output even in deaf signers using a gestural language such as American sign language or British sign language. Thus, the neural substrates underlying syntax are specialized for the functional role that syntax plays in language irrespective of its form in language. Function trumps form!

## 5.2.2    Problems with Grammar: Wernicke's Aphasia

Recall that in Chap. 2 we talked about the language output in Wernicke's aphasia. It was described as fluent and well-articulated, containing grammatical words and endings, and seemingly empty of semantic content. It certainly does not seem to be agrammatic, suggesting a real dichotomy in the syntactic abilities of Broca's and Wernicke's aphasia. Even the short samples we described in Chap. 2, repeated below, are replete with the use of grammatical free words, 'and', 'it', 'to' and the use of grammatical markers, e.g. 'saw', 'was', 'knew', wanted', and 'had'.

A4. 'I saw it and it was so that I knew it to be that it was'

A5. 'Every tay (day) I saw spalika when it had to'

Perhaps this means that syntactic deficits in Wernicke's aphasia are different in kind from those seen in agrammatism in Broca's aphasia. There is a problem, however. Now that we know that in agrammatism, neither free grammatical words nor bound grammatical markers are lost, but rather free grammatical words may be omitted and bound grammatical markers may be substituted for each other, what does it mean when we look at the language production in Wernicke's aphasia? Grammatical words appear as do grammatical markers on words. In the absence of knowing the meaning of what is being said, however, it is hard to say whether productions contain grammatical errors. For example, is the phrase in A4.,'…I knew it to be that it was' grammatical? And what would be the syntactic error or errors? What about 'every tay (day) when it had to' in A5. Are there grammatical errors here too?

It is not just the difficulty of interpreting the meaning of a sentence, but also its fluency that could trick us into thinking that all is well in the syntactic arena in Wernicke's aphasia. Does the fluency with which language is produced belie a grammatical impairment? That is, language comes out so smoothly with such clear articulation that it seems ok, just difficult to understand.

There may be a way out of this quandary. What we need to do is narrow down the set of possibilities of what can be said by controlling the context, the possible vocabulary, and the likelihood of a particular response. That way, we know what should be said and then can determine whether there are grammatical errors when speaking. One method is to present a drawings or a series of drawings for description. Analysis of the responses will allow us to determine the extent to which the speaker uses or omits free grammatical words and uses correctly or substitutes bound grammatical markers. For example, Fig. 5.2 elicited 'the boy gives the girl a book' or 'the boy is giving the girl a book'. If I now present Fig. 5.3 which shows the drawing of the boy giving the girl a book followed by a drawing of the girl now holding the book with a smile on her face, a likely description of this second picture would be 'the boy gave the girl a book'. Note that in contrast to Fig. 5.2 which elicited the present tense of the verb, Fig. 5.3 elicits the past tense of the verb to designate the completed action. It is possible to construct a set of pictures like these which elicit not only simple sentences like 'the boy is giving the girl a book', but more complex sentences, like 'the girl was given a book by the boy' or 'it is the boy who gave the girl a book'.

**Fig. 5.3** Example of a stimulus created to elicit the past tense of the verb 'give'. The drawing on the left shows a boy giving the girl a book. The drawing on the right, to be described by the subject, shows the completed action of the boy giving the girl a book

**Table 5.1** Types of Sentence Production Errors

| Features of syntax production | Performance |
| --- | --- |
| Sentences | Simplified |
| Occurrence of errors | Variable |
| Free-standing grammatical words | Omitted |
| Bound (attached) grammatical markers | Substituted |

It turns out that those with Wernicke's aphasia do have problems with producing syntactically correct sentences. This pattern of results is surprisingly similar to the pattern in Broca's aphasia (see Table 5.1). These similarities between the two aphasia groups emerge not just for English speakers but also for speakers of other languages such as German, Italian, Japanese, Finnish, and Dutch, all of which have a richer system of grammatical markers than English. Just like agrammatics, those with Wernicke's aphasia show variability in their performance. Sometimes they produce the correct forms. Other times they omit or use the wrong grammatical words and substitute wrong grammatical endings. Both Broca's and Wernicke's aphasics show the same order of difficulty in producing different sentence types, and they make more errors as the sentences are more complex syntactically. For example, sentences like 'the girl was given the book by the boy' are not only more difficult than 'the boy gives the girl a book' but more syntactic errors are produced.

However, there is one important difference that emerges between the two groups. Those with Broca's aphasia make consistently more syntactic errors than those with Wernicke's aphasia. We will return to this issue later in this chapter, where the quantity of errors but not the pattern of errors will provide us a potential explanation for the underlying nature of syntactic deficits in aphasia. Nonetheless, these findings are a hint that Broca's and Wernicke's aphasics have a common syntactic impairment despite their differing neural substrates and clinical picture.

Before we consider this possibility, however, we need to consider the important role that syntax plays in language comprehension. Here again, the clinical descriptions we presented earlier of Broca's and Wernicke's aphasia would suggest a dichotomy. Those with Broca's aphasia are said to have good auditory comprehension in contrast to those with Wernicke's aphasia who present with severe auditory comprehension deficits. Is this dichotomy real?

## 5.3    Problems with Grammar in Aphasia: Sentence Comprehension

Why do you need syntax to understand language? After all, if I say the sentence, 'Tom ate a hamburger', it would be easy to understand the meaning of the sentence. Here, we know that in the real world, boys eat hamburgers, but hamburgers do not, as a rule, eat boys. However, there are multiple instances in which information about the world or your expectations about what is going to happen cannot be relied on to understand the meaning of a sentence. 'The lion killed the tiger' is one such example. You have to know the syntax of the sentence to know who did what to whom and which animal is no longer with us. And it is not just word order in English that tells you who is the 'agent', the doer of the action, and who is the 'recipient', the receiver, of the action. It is the particular syntactic structure of the sentence that provides the information. Thus, in the sentence 'the tiger was killed by the lion', 'the tiger' is still the one in this scenario who is dead.

Perhaps it is the case that the reportedly good comprehension in Broca's aphasia reflects the ability to use knowledge about the real world and context to understand the meaning of sentences. This raises an even more important issue. Does the agrammatism seen in Broca's aphasia reflect a problem in speech output or does it reflect a 'deeper' deficit – a syntactic disorder that affects not only speaking but also comprehending? A syntactic disorder that affects both speaking and understanding would suggest that there is a common neural substrate for processing the syntactic properties of language just as we have shown for the speech component and for the lexical component.

And what about Wernicke's aphasia? Since we found that the agrammatic properties shown in Broca's aphasia are also seen in Wernicke's aphasia, whatever difficulties we find in Broca's aphasia may also arise in Wernicke's aphasia. Finding similar deficits in both groups would suggest that there is a broadly distributed neural substrate underlying syntax.

**Table 5.2**  Different Syntactic Structures used to Test Sentence Comprehension

| Sentences with different syntactic structures | Syntactic type |
|---|---|
| 1. The lion killed the tiger. | Active declarative |
| 2. The tiger was killed by the lion. | Passive |
| 3. It was the lion that killed the tiger. | Cleft subject |
| 4. It was the tiger that the lion killed. | Cleft object |
| 5. The lion killed the tiger and chased away the wolf. | Conjoined |
| 6. The lion that killed the tiger chased away the wolf. | Subject-object relative |
| 7. The tiger that the lion killed chased away the wolf. | Object-subject relative |

One way to test the comprehension of syntax is to present sentences to subjects and ask them to point to the correct picture from a set of differing scenarios. Another method is to present sentences, some of which are grammatical and others of which are ungrammatical, and ask the participant to make a grammaticality judgment by indicating whether the sentence they hear sounds ok or does not. Here, comprehension of meaning is not being directly tested. Rather the grammaticality judgment is based on how the sentence 'sounds' to the listener. For example, does the sentence 'he went to school in September' sound ok to you? What about 'he went school in September'?

The sentences presented in both comprehension and grammaticality judgment tasks typically cover a wide range of different syntactic structures Examples are shown in Table 5.2. It is not necessary for you to know the names of the individual syntactic structures, shown in the second column of the table. However, if you look at the examples, your intuitions are likely that some sentences are easier to understand than others. This is borne out by the performance of those with Broca's and Wernicke's aphasia. Sentences in which the subject of the sentence precedes the object (sentences 1, 3, 5, 6) are easier to understand than those in which the object of the sentence precedes the subject (sentences 2, 4, 7). Complex sentences like relative clause sentences in which a phrase is embedded in the sentence (sentences 6 and 7) are particularly difficult.

These results show that just like syntactic deficits in production, Broca's and Wernicke's aphasia share similar patterns of performance in comprehending and making grammaticality judgments about a range of syntactic structures (Wilson and Saygin 2004). Those with Broca's aphasia not only show agrammatism in their speech output, but they also show impairments in understanding sentences whose meaning depends solely on syntactic structure. When they cannot depend on context and what they know about the world, their comprehension of language is compromised. Similarly, those with Wernicke's aphasia show impairments in the comprehension of the syntactic properties of language, and the pattern of their performance mirrors that of Broca's aphasia.

There is one critical difference, however, between the two groups. Just as we saw a difference between Broca's and Wernicke's aphasia in the *severity* of the syntactic deficit in production, we find a difference in the *severity* of the syntactic deficit in comprehension. Wernicke's make substantially more errors than do Broca's.

But wait. These findings suggest that there is a dissociation between the two groups in their syntactic abilities with greater output deficits in Broca's aphasia than in Wernicke's aphasia and greater comprehension deficits in Wernicke's than in Broca's aphasia. Doesn't this sound (hint and pun intended) familiar? Remember in Chap. 3, we saw that there were greater errors in the production of sounds in Broca's aphasia compared to Wernicke's aphasia, and there were greater errors in the perception of sounds in Wernicke's aphasia compared to Broca's aphasia. Now, the same pattern is found in syntax; there are more syntactic production errors in Broca's aphasia compared to Wernicke's aphasia, and there are more syntactic comprehension errors in Wernicke's aphasia compared to Broca's aphasia. Thus, the severity of syntactic impairments mirrors those of the severity of sound processing impairments.

Taken together, the pattern of breakdown and the number of errors in production and comprehension provide us with important clues as to the neural architecture underlying syntax and the nature of the breakdown Broca's and Wernicke's aphasia. Here we go!

## 5.4   Following the Clues: What Goes Wrong with Syntax in Aphasia

As we described in Chap. 2, each component of grammar has its own functional architecture. After all, the representations that tell us what the sounds of language are cannot be the same as the representations for the meanings of words. The representations for sounds are based on their acoustic and articulatory properties whereas the representations for meanings are based on their semantic and conceptual properties. Nonetheless, sounds, words, and meanings each have a network-like architecture with connections within the network that influence its activity. It is the connection strengths which define the relationships between sounds, words, and meaning and the patterns of errors that occur in aphasia.

The architecture of syntax is more complicated. Think about it – what makes syntax, syntax. It is strings of words and the relationships between them as they unfold that define syntax. For example, 'the' plays a syntactic role in English only in relation to the words that follow it; its function grammatically results from the linguistic environment in which it occurs. Together, 'the' followed by 'boy' marks a noun phrase which plays a particular role in a sentence such as an agent (the doer of the action) in 'the boy likes the girl', or a patient (the receiver of the action) in 'the girl likes the boy'. So the functional architecture of syntax has to include a means of building structure over time.

At the same time, as we have discussed earlier, grammatical words and markers also have their own intrinsic meanings. 'The' in the phrase 'the boy' refers to a particular boy, whereas 'a' in the phrase 'a boy' refers to an unknown or unspecified boy. But both 'the' and 'a' have the same syntactic function which is to modify 'boy'. Similarly, markers that signify tense on a verb are added to the stem of the verb. Each has its own meaning; '-ed' is the past tense marker for the verb 'walk<u>ed</u>', whereas '-s' is the present tense marker, 'walk<u>s</u>'. Thus, the functional architecture of syntax needs to have both a linear component which predicts the occurrence of the words in their context as well as a non-linear component which groups the words into categories representing their syntactic function.

You might be wondering how the linear component is going to work. Let's assume a sentence starts with 'the'. There are thousands of words that can follow 'the' in English. However, there are limitations too. There can be, 'the boy', the dog', 'the rock', 'the building', and 'the can opener' as well as 'the nice boy', 'the big dog', 'the hard rock', 'the 'tall building' and 'the broken can opener'. However, there cannot be 'the walked', 'the the', 'the of', 'the in'. And there are words that may occur less frequently following 'the', but are still possibilities. For example, 'the very' is fine if following 'very' is an adjective such as 'the very nice' in 'the very nice boy'. This means that there are different *probabilities* about what types of words and also what specific words can follow each other in the syntactic component. Such *probabilities* generalize over all of the words that are likely to follow 'the'. There would be a high probability for the set of words that are nouns in English to follow 'the' (even though the network might not literally have a label for nouns), and a much lower probability for the set of adverbs like 'very' to follow 'the'.

In this way, the words of a language are organized into categories based on the syntactic roles they can play. These relationships and their probabilities can be captured by the connection strengths of the different nodes representing the grammatical categories in English. The set of nouns would have stronger connections to the set of determiners like 'the' and 'a' than to the set of adverbs like 'very'. As a result, the words in a sentence are not a simple linear string with equal probabilities between the words. Rather, the words in a sentence build up a structure based on how likely that set of words is found in the linguistic environment and also how likely they semantically cohere. Thus, the words in the sentence 'the the likes boy girl' would have weak connection strengths to each other since they have a minimal probability of occurring next to each other in a sentence context. And even though the sentences 'the boy likes the girl' and 'the rock likes the building' are both grammatical sentences, the first sentence has an overall higher probability of occurrence and thus greater connection strength among the words in the sentence than the second sentence. This is because animate folks can like each other, whereas inanimate objects like 'rock' do not emote, and hence they have a low probability of liking anything including buildings.

One last general point. The structure of a sentence is not based solely on the strength of connections between *adjacent* words. Rather, the ultimate syntactic structure of a sentence takes into account all of the words that preceded it. The full linguistic context must be included as the structure of the sentence is built, including even words that are distant from each other. For example, consider the verb 'walks' in the sentence, 'the boy who the girls like walks to school'. 'Walks' agrees with its subject noun phrase 'the boy' even though 'the boy' and 'walks' are not next to each other in the sentence.

It may be easier to conceptualize how the structure of a sentence is built with a simple example. Let's take the sentence, 'the boy walks a dog'. Figure 5.4 shows that each word is produced in sequence in time, one after the other. However, as shown by how the words are grouped in Fig. 5.4, each word also belongs to a particular syntactic category based on the syntactic function it plays in a sentence. Thus, 'a' and 'the' are grouped together in the same syntactic category since they are both determiners in English. The nouns of English group together as do the verbs. In this way, members of the same syntactic category are closely connected in the network, and hence more likely to compete with each other. Importantly, the organization of words into syntactic categories provides an accounting of the pattern of substitution errors that occur in aphasia. The substitution of one word for another typically respects the grammatical category it belongs to. Determiners substitute for determiners, nouns substitute for nouns, verbs for verbs, and so on.

Note that the tense markers are also grouped together but are linked directly to the verb. It may seem strange that they are linked but also grouped separately from the verb. Consider what we saw in the grammatical errors produced in both Broca's and Wernicke's aphasia. Grammatical endings like the ones marking tense were substituted for each other suggesting that they are in some sense independent units from the verb stem. Nonetheless, we also know from the pattern of errors in aphasia that grammatical endings are never produced in isolation, stripped from the verb stem. Neither are they added to the wrong part of speech. For example, a tense marker '–ed' is not added to a noun such as '*boyed'. Thus, a verb stem and its tense endings are inextricably tied together.

There is another way to capture the same set of findings for the substitution of grammatical endings for each other. Verbs and their endings may be indissoluble wholes and errors like 'walks' for 'walked' could reflect the substitution of one whole word verb form for another with a different tense of the same verb, such as the present tense of the verb 'walks', substituting for the past tense 'walked'.

At this point, it is not clear which is the correct way to represent these results. Are words quasi-independent units, stored separately from verb stems, or are words and their grammatical endings indissoluble wholes. Were we to take the second interpretation, we would need to have for every verb, a listing of the verb plus all of its possible tenses, e.g. walk, walks, walked, walking, talk, talks, talked, talking, jog, jogs, jogged, jogging, and so on. This would be required for all of the verbs of the language, in this case English, substantially increasing the number of possible verb forms to be held in memory.

I favor the first possibility (shown in Fig. 5.4) where a verb is linked but represented separately from its tense marker, and one tense marker substitutes for another tense marker. You may prefer the second. No matter, since for our purposes in both cases either architecture captures the error pattern in aphasia.

Up to this point, we have said that words are produced one after the other, and they are grouped into categories based on the syntactic roles they can play (see Fig. 5.4). The next step is to show the architecture of the syntactic component and

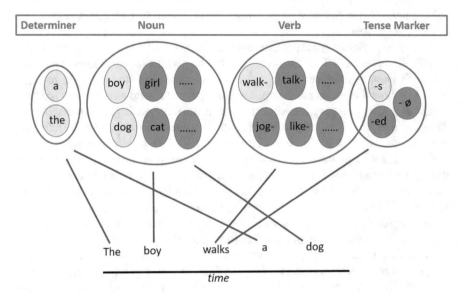

**Fig. 5.4** Syntactic category membership of the words in a sentence. Each word in the sentence is associated with its syntactic category membership as the sentence unfolds over time

build the syntactic structure of the sentence. As Fig. 5.5 shows, the syntactic structure of the sentence is built up based on the probability that particular syntactic categories occur together in the sentence, taking into account the probability of the syntactic category based on <u>all</u> of the input that precedes it.

Returning to our example sentence, 'the boy walks a dog', the structure of the sentence is built upon a set of probabilities: the probability that a noun, such as 'boy' follows an article as in 'the boy' forming a syntactic unit, a noun phrase, 'the boy'; the probability that a verb follows a noun phrase 'the boy walks'; the probability that an article 'a' follows a noun phrase and verb, in 'the boy walks a'; and the probability that a noun follows an article, again forming a noun phrase, 'the boy walks a dog'.

You now have the facts about the syntactic deficits in Broca's and Wernicke's aphasia and the network architecture of the syntactic component. Let's now put them together.

## 5.5    The End of the Story

Despite their different clinical picture and different neural loci of their lesions, Broca's and Wernicke's aphasia show similar patterns of performance in the production and comprehension of syntax. There are three take-aways from these findings. The first is that the syntactic component is broadly distributed in the brain recruiting multiple brain areas. The second is that the two groups have a common underlying deficit. Critically, the third take-away is that performance is variable. Sometimes

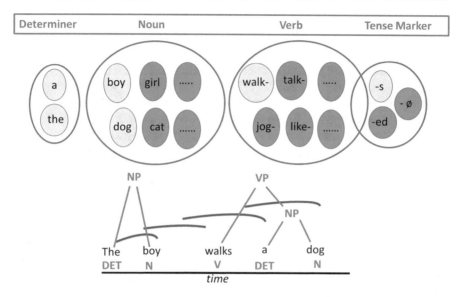

**Fig. 5.5** Architecture of the syntactic component. As the words in the sentence unfold over time, they are assigned to their syntactic category membership. The syntactic structure of the sentence built based on the probabilities that particular syntactic categories occur together. The red line shows the groupings over the preceding input and the probability they co-occur in the sentence (See also legend for Fig. 5.4)

productions are correct; other times they are not. However, importantly, syntactic errors are not random. Grammatical words and grammatical endings are substituted for each other and increased errors occur as grammatical complexity increases. This pattern can be captured by a network whose architecture is spared, but one which is operating in noise. As we have described in Chap. 2, a noisy network affects the efficiency with which the network operates leading to increased variability. It also lessens the differences between items that share syntactic properties resulting in a pattern of substitution errors where the wrong grammatical marker or the wrong grammatical ending is accessed.

A noisy network also affects the sequencing of words that ultimately give rise to syntactic structures. Context builds structure based on the probabilities of occurrence of grammatical categories as they unfold over time. The higher the probability that syntactic strings occur together, the more tightly bound they are to each other and the less vulnerable they are in a noisy system than are low probability sequences. Similarly, the grammatical structure of simple sentences is of higher probability than that of complex sentences and hence they are less vulnerable in a noisy system.

Other factors may also contribute to the vulnerability of complex sentences in both speaking and understanding. Complex sentences are typically longer than are simple sentences requiring more memory to process them. Additionally, complex sentences tend to be more semantically complex since they require the integration of semantic information over embedded phrases thus requiring greater processing resources than do simple sentences.

### 5.5.1   Vive la Difference

While the patterns of syntactic errors are shared in Broca's and Wernicke's aphasia, as we described earlier the numbers of errors differed. There were more syntactic errors in production in Broca's aphasia compared to Wernicke's aphasia. Contrariwise, there were more syntactic comprehension errors in Wernicke's aphasia than in Broca's aphasia. These findings are similar to what we saw in Chap. 3 when comparing the differences in speech production errors and speech perception errors in Broca's and Wernicke's aphasia. Recall that Broca's aphasics made more speech production errors than did Wernicke's, and Wernicke's aphasics made more speech perception errors than did Broca's. Here, we propose that the explanation for the differences between Broca's and Wernicke's aphasia in the frequency of errors in the production and comprehension of syntax is fundamentally the same as that found in speech production and speech perception and reflects the functional specialization in frontal and temporal areas of the brain. Frontal areas are specialized for the planning and implementation of speech, hence speaking, whereas temporal areas are specialized not only for mapping the acoustic signal on to the sounds of language but also for mapping words to their meanings, hence understanding. Thus, it is not surprising that differences emerge in syntactic processing with frontal lesions associated with Broca's aphasia resulting in greater difficulties in syntactic output and temporal lesions typically associated with Wernicke's aphasia resulting in greater difficulties in syntactic input.

The specialization of frontal areas for planning and implementation of speech may also play another important role in the character of the syntactic production deficit that we see in Broca's aphasia. Recall that clinically, English-speaking Broca's aphasics often omit free-standing grammatical words, whereas those with Wernicke's aphasia typically retain them. It turns out that free-standing grammatical words such as 'the' and 'of' are typically shorter in duration and lower in amplitude (loudness) when produced than are the content words of a sentence such as nouns and verbs. These free-standing grammatical words may thus be 'collateral damage' in the context of a deficit in planning and articulating speech, exacerbating the syntactic deficit in Broca's aphasia.

### References

Bock, J. K. (1986). Syntactic persistence in language production. Cognitive Psychology, 18(3), 355–387.

Garrett, M., Bever, T., & Fodor, J. (1966). The active use of grammar in speech perception. Perception & Psychophysics, 1(1), 30–32.

Wilson, S. M., & Saygin, A. P. (2004). Grammaticality judgment in aphasia: Deficits are not specific to syntactic structures, aphasic syndromes, or lesion sites. Journal of Cognitive Neuroscience, 16(2), 238–252.

# Readings of Interest

Dell, G. S., Chang, F., & Griffin, Z. M. (1999). Connectionist models of language production: Lexical access and grammatical encoding. Cognitive Science, 23(4), 517–542.

Dick, F., Bates, E., Wulfeck, B., Utman, J. A., Dronkers, N., & Gernsbacher, M. A. (2001). Language deficits, localization, and grammar: evidence for a distributive model of language breakdown in aphasic patients and neurologically intact individuals. Psychological Review, 108(4), 759.

Elman, J. L. (1991). Distributed representations, simple recurrent networks, and grammatical structure. Machine Learning, 7(2–3), 195–225.

Menn, L., and Obler, L. K., (Eds.). (1990). Agrammatic aphasia: A cross-language narrative sourcebook (Vols. 1–3). John Benjamins Publishing.

# Why Two Hemispheres: The Role of the Right Hemisphere in Language

Up until now, we have focused solely on the role of the left hemisphere in language. Does this imply that the right hemisphere is moribund with respect to language? As we will see, the answer is complex. Indeed, perhaps more than any other aspect of the functional organization of language in the brain, the right hemisphere remains an enigma. Some evidence suggests that the right hemisphere has similar abilities to the left hemisphere. Other evidence suggests that the right hemisphere is a 'poor step-child' of the left hemisphere, having some language abilities but limited in scope and linguistic richness. Still other evidence suggests that the right hemisphere plays a different but essential role in language communication.

Let us begin by looking at the effects of right hemisphere brain injury on language. If it has language functions similar to those of the left hemisphere, then right hemisphere damage should result in aphasia, and just as we saw with damage to the left hemisphere, we should see different patterns of language impairment as a function of the area of the brain that is damaged.

## 6.1 Aphasia and the Right Hemisphere

Recall that persons with aphasia show deficits in different aspects of language production and comprehension affecting speech sounds, words, meaning and/or syntax. The question is whether the right hemisphere has the same language functions as the left hemisphere. The results are clear-cut. Damage to the right hemisphere typically does <u>not</u> produce an aphasia, the exception being a small percent of individuals, typically left-handers, who have crossed dominance meaning that language is represented in the right hemisphere. Although some mild impairments surface when looking at particular aspects of language, right hemisphere damage does not produce the constellation of impairments we have seen in the aphasia syndromes pursuant to left hemisphere injury. Let's look in a little more detail at the language performance of those with right hemisphere lesions starting with the sounds of language and then turning to words, meaning, and syntax.

© Springer Nature Switzerland AG 2022
S. E. Blumstein, *When Words Betray Us*,
https://doi.org/10.1007/978-3-030-95848-0_6

Patients with right hemisphere damage process the sounds of language similarly to those without brain injury. They rarely, if ever, make sound substitutions while speaking, and they produce the sounds of language normally. They respond correctly when asked whether pairs of similar sounding nonsense syllables like 'pa' and 'ba' or words like 'pat' and 'bat' are the same or different. Thus, they do not show deficits in perceiving the sounds of language. However, some patients do have problems with producing *prosody*, the 'melody of speech', especially when they use it to convey emotions like 'happy', 'sad', 'angry'. We will return to this aspect of speech a little later in the chapter when we talk about language communication more generally.

Right hemisphere patients can map sounds to words and words to meanings. Generally, they can name pictures of words, and they can point to the correct picture or object when the word is presented auditorily. However, some errors do occur when they are asked to select a word from an array in which one of the words is similar in meaning to it. For example, asked to point to an 'apple' given four pictures including an apple, a pear, a shoe, and a ball, they might incorrectly point to 'pear'. Although such errors are not common, this pattern of errors is similar to what we saw for those with Broca's and Wernicke's aphasia and suggests that the architecture of the lexicon is similar to that we described for the left hemisphere.

As for syntax, unlike left hemisphere patients with aphasia, right hemisphere patients respond correctly when given sentences like 'the lion was killed by the tiger' and asked 'which animal is dead?' As described earlier, in order to respond correctly to such sentences, it is not possible to depend on real world probabilities; after all, lions can kill tigers and tigers can kill lions. To understand this sentence, it is necessary to process the syntactic 'rules' of English.

The absence of aphasia after right hemisphere brain injury suggests that the right hemisphere is not playing much of a role in processing language. End of story? Not quite. Let's step back for a second and consider what language is. It is more than saying and understanding words and sentences. We use language to communicate – to express our feelings and views, and to interact with the people around us. To do so, language has many 'devices' which add richness to its use. We use metaphor, humor, irony; we infer intentions and information beyond what is actually given; we make assumptions about others' beliefs taking for granted or presupposing certain truths. All of that shapes our choice of words and sentences as speakers and our comprehension as listeners. And here, the right hemisphere plays a crucial role. Namely, in contrast to the left hemisphere which is the consummate 'linguist' – concerned with the structure of the pieces of language, its sounds, words, meanings, and syntax as we speak and understand – the right hemisphere appears to be the 'social' half of the brain, sensitive to the context in which language is utilized as we communicate.

## 6.2    Beyond Sounds, Words, and Syntax

Interacting with a patient with right hemisphere injury can have its challenges. Such individuals often have difficulty maintaining a conversation because they fail to stay on topic or else they focus on some aspect of the conversation, providing excessive detail. They may make incorrect or unusual interpretations as they converse with others leading to misunderstandings or the sense that their language is a bit 'off'. For example, engaged in a discussion of the strengths and weaknesses of individual candidates for an upcoming election, someone with right hemisphere injury might interject 'only registered people can vote'. The answer is certainly about elections, but it is not germane to the conversation at hand.

There are other difficulties in interpreting language. All tie in in some way or other with taking language literally, and, as a consequence, seeing language as a message rather than as an interactive conversation which requires at times changing perspective or thinking from the vantage point of others. Let's start with the most obvious way language may be taken literally by considering the interpretation of metaphors. Some metaphors are frozen, like 'He kicked the bucket'. Here, by convention, the whole phrase has a totally different meaning than the interpretation of the individual words. Used metaphorically, 'kick the bucket' means to die, not that someone kicked a bucket. Not all metaphors are frozen. Words or phrases may be interpreted figuratively rather than literally. For example, 'he is a cool guy' refers to his being hip, not being in need of a sweater.

Right hemisphere damaged individuals typically prefer the literal rather than the metaphorical meaning of metaphors. 'Kick the bucket' is interpreted as 'letting the bucket fly with your foot'. They even have difficulty interpreting a word like 'warm' figuratively in 'she was a warm person' or understanding the figurative meaning of a sentence like 'trying to come up with the word, she felt like her head was full of cotton'. In both cases, those with right hemisphere damage typically give a literal interpretation – they interpret 'warm' as temperature in the first sentence, and they interpret 'head full of cotton' in the second sentence as an individual who has cotton between her ears.

Interpreting language literally can have humorous if not annoying consequences. Let's consider the following scenario. At a gathering of a family at the dining room table for a delicious Thanksgiving dinner, grandma, a small and delicate elderly lady, says, 'it's cold in here!' 'Yes, it is' replies her right hemisphere brain-injured nephew while continuing to eat. Was grandma looking for confirmation of the temperature of the room? No. She was giving an implicit command – 'please, close the window!' As the dinner progresses, grandma says, 'can you please pass the salt.' The response: 'I sure can' and again her nephew continues eating. What's going on here? Her nephew is interpreting grandma's statement as a remark about the salt, and he is not interpreting her statement as a request for an action.

Displaying a rigidity or inflexibility in interpreting language input can also influence the ability to 'switch' from an original interpretation of a preceding phrase or sentence. Those with right hemisphere damage have shown difficulties in changing an original inference they make when given further information that requires reinterpretation. If I say to you, 'I am so bored with this chapter', you would infer that I am reading the chapter and find it beyond not interesting. However, if I follow this sentence with 'If I don't finish it, my editor will be very annoyed', you will not interpret the original sentence to mean that I am bored *reading* the chapter. Rather, you would re-interpret the first sentence to mean that I am bored because I have been trying to *write* a chapter that will not end. Even after hearing the second sentence, those with right hemisphere damage fail to alter their original inference and continue to interpret the first sentence to mean that the speaker is bored reading the book.

There are other problems which right hemisphere patients have that affect their ability to be good communicative partners. They have difficulty understanding or appreciating humor, and most importantly, in integrating a punch line into a preceding narrative. What makes something funny? Typically, although the ending may fit in with the prior setup, there is a 'surprise' or unexpected denouement. So let's see how funny you find this joke: 'A duck comes into a bar with his friend, the gorilla, and the two order drinks. They get into an argument about who will pay and the gorilla says, 'the bill is on you'. The duck replies as he points to his face while leaving the bar smiling, 'it sure is'.' Right hemisphere patients would likely not see the humor in this (you may not either!). They fail to see that at the start of the story 'bill' refers to money owed for the drinks, but by the end of the story, there is a surprise with the word 'bill' referring to the duck's beak.

Is there a common explanation or underlying basis for many of these difficulties? Surely, we don't believe that there are separate modules for jokes, for inferences, for literal and figurative interpretations of language. There are several hypotheses for drawing connections between several of these aspects of language.

One possibility is that those with right hemisphere brain injury have a subtle semantic deficit. As we discussed in Chaps. 2 and 4, words and their meanings are not silos, but are rather organized in a network of connections. Thus, activating a word like 'dog', not only activates its meaning representation for 'dog', a four-legged hairy canine that has a tail and ears and can bark, but also a network of other words sharing its meaning or associated with it in some way or other. For example, dogs chase cats, they howl, they bite. Some of these connections are closely related to 'dog', others are more distant. Normally, close relationships are more highly activated, whereas more distant ones are more weakly activated. This appears not to be the case for the right hemisphere where activation of more distant meanings of words and their features appears to be as highly activated as close relationships. The result is that the lexical and semantic network of the right hemisphere is *coarsely* tuned, allowing more disparate relationships to be activated and used, in contrast to the left hemisphere network which is finely tuned, maintaining clear-cut distinctions among words and meanings in the network (Jung-Beeman 2005; Tompkins 2008). As a result, damage to the right hemisphere affects the ability of the network to select among the words and meanings that are part of the network. The consequence in the real world is difficulty in integrating lexical and semantic information in discourse leading to problems in staying on message in conversations, comprehending non-literal language such as idioms and metaphors, and drawing connections required when making inferences.

There is another possibility which was hinted at a bit earlier in this section. Those who have sustained right hemisphere brain injury appear to have difficulty in understanding that there are different perspectives or interpretations possible in a given social situation. In other words, they appear to be 'unto themselves' with a world-view that revolves around themselves, failing to grasp that others have thoughts, beliefs, and intentions which guide their behavior and which may be different from their own. As a result, their interpretation of language appears rigid or literal, as we saw in interpreting metaphors or the wants or desires of others, and inferring that further contextual information may require a change in an original 'belief' or interpretation.

Supporting the view that right hemisphere damage affects their 'world view' is the occurrence of another impairment. Those with right hemisphere brain injury often have difficulty recognizing the different emotions of others such as whether someone is happy, sad, or angry. These difficulties emerge whether listening to the prosody (the melody) of language produced by others or looking at their facial expressions. There are some with right hemisphere brain injury who also have deficits in producing different emotions in their own speech. Consider the 'social' effects of such difficulties. How does one react to someone who does not respond to you appropriately when you are sad or angry or who does not appear to share your joy when you are happy?

The ability to understand that others have mental states that may differ from your own is called a *theory of mind*. Children as young as 4 years old show that they have developed a theory of mind. They understand that other people have their own thoughts, beliefs, and intentions; that these may be different from their own; and that the beliefs of others may in fact be incorrect. If right hemisphere patients do indeed have a deficit in theory of mind then what appears to be an impairment in the *pragmatics* of language (the use of language in social context) may be a consequence of a cognitive deficit and not a language deficit. That is, a cognitive impairment manifests itself in and has consequences for the use of language. Whether the basis for this 'social' impairment with right hemisphere brain injury has its roots in a language deficit, a cognitive deficit, or both is unclear. However, the fact remains that damage to the right hemisphere results in a deficit in the *use* of language.

## 6.3    The Two Sides of the Brain Need Each Other

Taken together, the effects of right brain injury show that the two 'brains', the left hemisphere and the right hemisphere, need each other for normal language processing – the left hemisphere to construct and/or analyze the pieces of language and the right hemisphere to use language appropriately in its social context. And yet, although right hemisphere damage does not produce a frank aphasia nor do we see the emergence of aphasia syndromes as happens with those individuals with left hemisphere damage, injury to the right brain does lead to some mild language difficulties especially around the meanings and interpretations of words and sentences. So have we missed something? Is it possible that the right hemisphere does have similar language functions as the left hemisphere, but the failure to show language (linguistic) deficits similar to those that emerge with left brain injury reflects the *dominance* of the left hemisphere? Language dominance could mean that because the left hemisphere is the 'controller' of language, it potentially inhibits the active participation of the right hemisphere in language functions. If this were the case, then the right hemisphere might have the *capacity* for language, but it is unable to show its stuff.

There is a simple way to test this possibility. That would be to look at what each hemisphere can do independent of the other. Wouldn't it be interesting to see the right hemisphere at work without the 'interference' of the left hemisphere? We can learn about the workings of each hemisphere if we look at what happens when the two hemispheres are split, separated from each other so that each one can function independently of the other.

## 6.4    Split-Brain Patients

How could one split the two hemispheres of the brain? As we described in Chap. 2, the brain is divided into two hemispheres, the left and right hemispheres. They are connected by the corpus callosum (also called the cerebral commissures), a band of white matter fibers, which transmit information from one hemisphere to the other. Splitting or cutting the cerebral commissures can 'separate' the information flow from one hemisphere to another (see Fig. 6.1).

As one might expect, this is not a procedure to be taken lightly. However, this procedure turned out to have great potential for a subset of people who had intractable seizures due to epilepsy, a disorder in which there is abnormal neural electrical activity. In these cases, seizures can occur 100s of time in a day, can travel from one hemisphere to another, cannot be controlled pharmacologically, and ultimately are life-threatening. A solution was sought to limit the seizure activity of these patients to only one hemisphere.

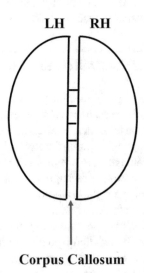

**LH     RH**

**Corpus Callosum**

**Fig. 6.1** Superior view (looking at the top) of the brain. The left hemisphere (LH) and the right hemisphere (RH) are connected to each other by the corpus callosum, white matter fibers that connect the two hemispheres. Splitting the corpus callosum prevents information from going from one hemisphere to the other hemisphere

A potential solution was found, based on the pioneering work of Roger Sperry in the 1950s who was interested in learning about the functions of the two hemispheres as well as the role of the corpus callosum. To do so, he and his colleagues cut the cerebral commissures in monkeys and in cats, and found, among other things, that there was no noticeable change in the animals' daily behavior. The animals could still perceive, act, and learn. Here then was a possible solution to the serious epileptic disorder of human patients.

Philip Vogel and Joseph Bogen performed split-brain surgery on humans in the 1960s. Importantly, there were no noticeable behavioral changes for most of these patients in their day to day activities and in their language. However, interacting with these patients on a daily basis cannot answer the question of whether the right hemisphere has the capacity for language independent of the left hemisphere since both hemispheres would be getting the linguistic information. To look at the 'behavior' of the two hemispheres separately requires using experimental methods that allow language information to be sent to only one hemisphere. This can be easily accomplished given the anatomy of the brain.

### 6.4.1  A Brief Interlude on Brain Neuroanatomy

Let's start with your hands. If you touch something with your right hand, the sensation innervates tactile (touch) areas of the *contralateral* (opposite) left hemisphere of your brain. This tactile information then flows to the right hemisphere by crossing the corpus callosum. In analogous fashion, if you touch something with your left hand, it activates tactile areas of the right hemisphere. Tactile information gets to the left hemisphere by crossing the corpus callosum.

Think about it. Tactile information on one hand directly innervates the contralateral hemisphere. Cutting the corpus callosum prevents information from one hemisphere from getting to the other hemisphere.

Let's now consider your eyes. Each eye has two visual fields, a left visual field and a right visual field in which visual information comes in from the outside world. Your left visual field projects from both eyes to your right hemisphere, and your right visual field projects from both eyes to your left hemisphere. Information crosses from one hemisphere to another via the corpus callosum. Here again, visual information can be isolated to one hemisphere by cutting the corpus callosum.

### 6.4.2  Language Processing in the Split Brain

A series of pioneering experimental studies with split-brain patients led by Michael Gazzaniga revealed critically important findings. Let's start with the tactile presentation of stimuli to the left or right hand. In these experiments, the split-brain individual is given an object in one hand. The object is out of view of the subject so naming can only be accomplished by touching the stimulus and not also by seeing it. The task is to name the object. For example, let's say the split-brain subject is asked to touch a ball out of sight behind a screen with his right hand. What do you think happens? He is able to name the ball because information from his right hand innervates the left hemisphere where language resides. What about when the subject touches and feels the ball in his left hand? Results show that he fails to name it. These findings tell us that the right hemisphere cannot say what the object is. Why? Because the tactile information is sent only to the right hemisphere which is unable to name objects. The tactile information cannot cross the corpus callosum to the left language hemisphere to accomplish the task.

Testing what the two hemispheres can do with visual presentation is a bit more complicated. An apparatus is needed to present stimuli to only one visual field. To do so requires that the subject fixate on a dot in the center of a screen. The stimuli are presented quickly (in milliseconds) either to the right or to the left of this fixation point so that the visual information can be presented only to one visual field and to the opposite hemisphere (see Fig. 6.2). When stimuli are shown to the right visual field (which projects to visual areas in the left hemisphere), the split-brain subject can name the object. When the stimuli are shown to the left visual field (which projects to the right hemisphere), the split-brain subject is unable to name the object – in fact, the subject typically responds not seeing anything. These findings show that the right hemisphere fails to process this linguistic information, and cutting the corpus callosum prevents it from being processed in the left (language) hemisphere.

Looks like that is the end of the story. These findings suggest that the right hemisphere has no capacity for language. However, the story is much more interesting. Further study showed that indeed the right hemisphere does have some capacity for language.

Even though the right hemisphere was unable to speak, behavioral studies showed that the right hemisphere is able to recognize some objects or written words when the task is to match that object or word from an array of objects or words. For example, shown the picture of a cup in the left hand or the left visual field, split-brain patients can pick out the cup from a set of objects including a glass, a fork, a screwdriver, and a cup. The right hemisphere has also shown an ability to read some (but not all) letters, numbers, and short words.

Being able to recognize individual words is important, but as you may remember, words are not silos, separated from other words, concepts, and meanings. Rather, they are a part of a semantic network, connected by meaning, function, use, and a range of parameters. So the question is, does the right hemisphere show evidence that words are stored in a network, and are the properties of that network similar to those of the left hemisphere? While there is not a lot of evidence, the findings are suggestive that, indeed, words accessed by the right hemisphere are organized in a semantic network, and that network is similar to that of the left hemisphere.

One of the most striking examples is the following. The word 'horse' was presented to the left visual field (hence right hemisphere) of a split-brain patient who was asked to draw a picture of the word (Gazzaniga 1985). He drew a horse. When asked what he saw, he replied 'nothing'. When asked to draw with his left hand what was presented, he drew a horse; and when asked what goes on it, he drew the picture of a saddle. Why is this a stunning example? Firstly, that he could draw a horse

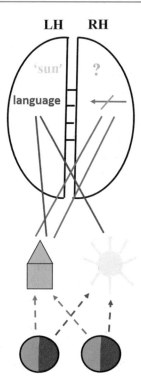

**Fig. 6.2** Presentation of visual stimuli to the right or left visual field of a split-brain subject for naming. When the image of the sun (in yellow) is presented to the right visual field of each eye (in red at the bottom of the figure), this information projects (red line) to the left language hemisphere (LH), and the split-brain patient correctly names it. When the image of the house (in green) is presented to the left visual field of each eye (in blue at the bottom of the figure), this information projects (blue line) to the right hemisphere (RH) but cannot cross the corpus callosum to the left hemisphere for naming. The patient reports seeing nothing. (See text)

showed that the right hemisphere was able to read the visual stimulus and understand what it represented. As to drawing the saddle, keep in mind that saddles are not an intrinsic part of horses. They are semantically related to horses, and are a part of the network of connections relating horses to objects associated with horses.

Similarly, shown a picture of a steaming cup of coffee in his left visual field, one of the split-brain patients could not name what he saw. However, with his left hand, he pointed to a card with the word 'hot' on it (Gazzaniga 1970). What this means is that the right hemisphere did see the steaming cup of coffee and processed a semantic property associated with it. 'Hot' is a characteristic feature of coffee, but it is not intrinsic to its meaning. After all, on a hot summer's day, you might order an iced coffee. My grandmother always put an ice cube in her hot coffee, and just like in the fairy tale of the three bears, it was neither too hot, nor too cold; it was just right!

As we have seen with aphasic patients, knowing words and their meanings is one thing, putting them together to form sentences is another. Here, the right hemisphere showed its limitations, displaying minimal ability to use syntax in understanding sentences. Shown pictures of scenes and required to match the scenes with a verbal description of the actions by the examiner, the right hemisphere of split-brain patients showed chance performance in understanding active versus passive sentences such as 'the boy kisses the girl' compared to 'the girl is kissed by the boy', understanding present versus future tenses of verbs in sentences such as 'he is driving the car' compared to 'he will drive the car', and understanding singular versus plural objects such as 'one book' compared to 'two books'.

In sum, the split-brain data suggest that there is some limited language capacity of the right hemisphere. The extent of this capacity varied across patients, with some showing a very primitive language ability and others showing greater capabilities. However, even at its best, the capabilities of the right hemisphere were never equivalent to those of the left hemisphere. Interestingly, the most language was seen in the right hemisphere when surgery occurred at earlier ages, suggesting that there may be a time-limited trajectory for the right hemisphere to acquire language.

The consequence of a developmental trajectory is that the extent to which the right hemisphere processes language will vary depending on when the two hemispheres are split. With an intact corpus callosum, the left hemisphere dominates over the right hemisphere, preventing it from realizing its capacity. When the corpus callosum is split early in development, the left hemisphere loses its ability to inhibit the right hemisphere's ability to process language. The earlier the split, the more language develops in the right hemisphere; the later the split, the less language displayed by the right hemisphere.

Split-brain research has given us a unique view of the potential of the right hemisphere to process language. However, there is an even more extreme approach we can take in determining the right hemisphere's capacity for language and the extent to which it is influenced by time. We can examine what happens to language when the entire left hemisphere is excised in children and in adults, leaving only the right hemisphere to process language.

## 6.5     A Radical Procedure

As you might expect, performing a *hemispherectomy*, the excision of an entire cerebral hemisphere, is done only under the most extreme circumstances. Consider the potential consequences of a left hemispherectomy. If the right hemisphere has limited language capacity and cannot compensate for the language skills of the left hemisphere, then performing a left hemispherectomy will have a devastating effect on language. That is exactly what happens for adults. Left hemispherectomy results in a severe aphasia and communication skills are limited, at best. This finding tells us that in the adult, the left hemisphere rules, and the right hemisphere, now presumably disinhibited from the control of the left hemisphere, is still unable to take on the job.

There is a different story for children who undergo left hemispherectomy leaving only the right hemisphere to process language. In most cases, after initial muteness and language difficulties, children quickly recover (in weeks or months), and their language appears to reach near normal levels. Only under close examination are deficits shown and they typically involve some residual articulatory problems and impairments in the comprehension of sentences that require reliance solely on syntax for understanding (see Table 5.2 for examples of these types of sentences). It appears then that the right hemisphere <u>does</u> have the capacity for language and shares the same (or similar) functions as the left hemisphere in processing the sounds, words, meanings, and syntax of language. This is good news as it suggests that the right hemisphere may take over language functions of the left hemisphere when the left brain is injured (for further discussion see Chap. 7).

However, there is sobering news as well. There is a developmental trajectory to language such that the language capacity of the right hemisphere is time-limited. The brains of children (up to around puberty) appear to be able to adapt to or compensate for left brain injury. However, from what we have seen, the right hemisphere does not appear to have the ability to do so in the adult brain.

## 6.6    It's Still a Puzzle

It is clear that the two sides of the brain work in tandem – the left hemisphere to construct the pieces of language and the right to use those pieces in the context of a social world. This would seem to be a simple but interesting division of function. Both aspects of language – the structural and the social – are critical and necessary for its normal use. But the story is a bit more complicated. Looking at the evidence from multiple vantage points (see also Chap. 8), it is unclear what linguistic functions are also the province of the right hemisphere. As we have seen, in young children, the two hemispheres appear to be *equipotential* or nearly so; that is, the two hemispheres have similar linguistic functions and each can process language. This capacity seems to emerge, however, only when the left hemisphere is completely 'deactivated' through its complete removal by hemispherectomy in children. When there is left brain injury in adults and when the two hemispheres are disconnected in split-brain patients, the right hemisphere shows only some language capacity, but it is limited and primitive.

We end this chapter unable to say whether the right hemisphere plays an active role in language processing in the adult brain. We will continue this discussion in Chap. 7 when we look at language recovery in aphasia and in Chap. 8 when we consider additional findings using approaches other than lesion studies to inform us about the brain and language.

## References

Jung-Beeman M. (2005). Bilateral brain processes for comprehending natural language. *Trends in Cognitive Science*, 9(11), 512–518. doi: https://doi.org/10.1016/j.tics.2005.09.009. Epub 2005 Oct 7. PMID: 16214387.

Gazzaniga, Michael S. (1970). The Bisected Brain. New York: Appleton-Century-Crofts.
Gazzaniga, M. (1985). The Social Brain. New York: Basic Books.
Tompkins CA. (2008). Theoretical considerations for understanding by adults with right hemi-sphere brain damage. *Perspectives on Neurophysiology and Neurogenic Speech and Language Disorders*, 18(2), 45–54. doi:https://doi.org/10.1044/nnsld18.2.45.

## Readings of Interest

Code, Chris. 1987. Language, Aphasia and the Right Hemisphere. New York: John Wiley & Sons.
Nebes, R.B. 1978. Direct examination of function in the right and left hemispheres. In Marcel Kinsbourne (ed.). Asymmetrical function of the brain. Cambridge University Press. Pp. 99–137.
Springer, S. P. and Deutsch, G. (1981). Left Brain, Right Brain. San Francisco: W.H. Freeman and Company.

# The Plastic Brain

**7**

In the previous chapters, we have looked at the devastating effects of brain injury on language. Although we have hinted at it, we have not yet considered in any detail the extent to which the brain is *plastic*; that is, we have not considered its ability to adapt to its injury. The degree to which this is possible has tremendous implications for recovery of language in aphasia. Think about it – if the brain is truly hard-wired with specific neural areas having discrete and dedicated functions, then damage to those areas would present a gloomy, if not disastrous picture for language recovery. However, if the brain is plastic such that other areas of the left hemisphere or even the right hemisphere have the capacity to 'take over' or at least accommodate to the neural injury, then the better the chances would be for regaining or at least improving upon impaired language functions.

When we look at the potential for plasticity in aphasia, we are looking at how a damaged system may compensate for the functions those neural areas play in language. This will be one topic we will explore in this chapter. But what happens if, rather than injury to the brain, the areas of the brain that are used for language are *deprived* of crucial input. Such is the case for those who are deaf or blind. Here, an inability to hear prevents auditory information from reaching those neural areas that process speech and this could have a cascading effect on the other components of language. Similarly, an inability to see prevents visual information from reaching those neural areas involved in reading. In such instances, what are the consequences for language? Is the brain plastic enough to accommodate to the loss of this input? And think about what it means when neural areas are deprived of input. Do those neural areas not utilized lie dormant or are they able to be 'repurposed' to play a functional role in language?

These are the questions we hope to answer in this chapter. But before we do, let's review some facts we do know about neural plasticity and language. Yes, the brain is plastic. As we learned in Chap. 6, both hemispheres of the brain appear to have the capacity for language. Brain injury to the left hemisphere and even its full removal (*hemispherectomy*) can result in a nearly full recovery of language.

© Springer Nature Switzerland AG 2022
S. E. Blumstein, *When Words Betray Us*,
https://doi.org/10.1007/978-3-030-95848-0_7

However, there is a serious caveat. Namely, there is a *critical period,* a time, during which the brain is most able to 'recover'. The greatest potential occurs from birth through childhood and plateaus around puberty.

Critical periods play a role not only in recovery of language after injury but also when learning a language – whether it be learning the first or second language, learning sign language (the communication system for the deaf), or learning Braille (the reading system used by the blind). Although children have a distinct advantage in learning language, it does not mean that there is no neural plasticity in the adult. Adults do learn new languages, and can learn sign language and Braille, but their linguistic abilities rarely reach the level of those who acquired language during the critical period. Those of you who are adults, reading this paragraph, do not despair. As adults, we are learning all of the time. We are constantly learning new words such as *twitter* and *blog*, and new slang phrases such as *my bad*, and we use them with facility and ease. But learning new words or phrases is not the same as learning a new language in its entirety – its sounds, producing or perceiving them like a native speaker, its rich vocabulary, its complex syntactic structures and ultimately the meanings conveyed.

Of interest, there are neural changes that show the effects of practice as we learn. Figure 7.1 shows an example of such changes (Posner and Raichle 1994, p. 127). In this neuroimaging experiment, subjects are given a noun and are asked to name a verb related to it. For example, given the noun 'dog', they might reply 'bark'; given the noun 'shoe', they might reply 'walk', given the noun 'car' they might reply 'drive'. The left column shows the neuroimaging results when subjects do the task for the first time. Note the activation in the left frontal (top and middle row) and temporal areas (middle row) and in the right cerebellum (bottom row). After practice, there is reduced activation in these areas. Indeed, it appears as though the brain has become dormant. Has it? Not at all. It turns out there are two systems which recruit different brain areas. When the word responses are practiced they become 'automatic', leading to a reduction of neural activity in the areas originally activated when the task had been done for the first time. Another system (not pictured) is activated when the words have become learned. Having two different systems has important consequences. It allows for the neural system to distinguish between the use of language 'creatively' such as generating a novel sentence to communicate an idea and the use of the overlearned 'habits' of language such as counting or saying the alphabet. This distinction between these two aspects of language is evident in aphasia where the ability to produce overlearned aspects of language may be relatively spared while the ability to generate novel sentences is impaired.

But we have digressed. It is time to return to the questions at hand. Let us begin by first examining the role plasticity plays in recovery of language in aphasia. We will then turn to neural plasticity and the role it plays in communication for those who are deaf and those who are blind.

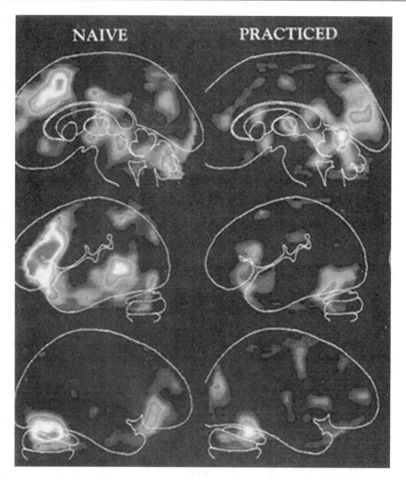

**Fig. 7.1** Activation patterns showing neural changes occurring when a task is new (naïve) and when it is learned (practiced) (see text). Three vertical slices are shown in each column with areas of activation displayed in red and yellow. The top two rows in each column show areas of the left hemisphere and the bottom row areas of the right hemisphere. Adapted from Posner, M. I. and Raichle, M. E. 1994. Images of Mind. Scientific American Library/Scientific American Books

## 7.1 Language Plasticity After Brain Injury

The effects on language of brain injury to the left hemisphere can vary extensively. For some, the aphasia resolves quickly, even within hours or days, and the 'nightmare' is over. For others, however, language recovery is slow and gradual. There is typically improvement, but the degree of recovery may vary, leaving many with a residual aphasia. That there is any recovery tells us that the brain is plastic and is able to accommodate or reorganize in some way to the brain injury. The question is how?

Language recovery is a gradual, evolutionary process and neural reorganization over time presumably underlies this course of recovery. Neuroimaging studies show that at early stages of language recovery, there is increased activation in the right hemisphere. These findings have been used as evidence that the right hemisphere is contributing to language recovery. However, there is evidence that challenges this conclusion. At later stages of recovery (around 6 months or later), there is a reduction of right hemisphere activity and an increase in activity in *perilesional* areas (areas around the lesion) of the left hemisphere. Indeed, those with the greatest language recovery show reduced right hemisphere activity and increased left hemisphere activity.

Like most straightforward descriptions, the story is actually more nuanced. After all, the devil is in the details. Let's begin first with the role of the right hemisphere, as it appears to be a player in the road to recovery. There is some good news and some bad news. Some evidence points to a beneficial role the right hemisphere plays in language recovery (good news), and other evidence suggests it is a spoiler, like a sore loser, playing a maladaptive role and getting in the way of the left hemisphere recovery process (bad news). Let's start with the good news.

We know that the left hemisphere is dominant for language. One scenario which we brought up in Chap. 6 is that under normal circumstances the left hemisphere dominates and thus inhibits the right hemisphere from playing an active role. Damage to the left hemisphere may upset this uneven 'balance of power', allowing the right hemisphere to contribute to language recovery. Indeed, several case studies have shown that after some recovery of language from a left hemisphere stroke, a second stroke in the *right* hemisphere worsens the aphasia deficit suggesting that the right hemisphere was responsible for the initial recovery of language. Indeed, a number of treatment studies have shown increased right hemisphere activation with improved language. These findings point to brain plasticity with a critical role played by the right hemisphere in language recovery.

Now for the bad news. There is some evidence that the right hemisphere actually gets in the way of the left hemisphere. Recall that the left hemisphere dominance may be diminished after left brain injury resulting in increased neural activity of the right hemisphere. Noninvasive brain stimulation studies using a method called repetitive transmagnetic stimulation (rTMS) have shown that *reducing* the brain activity of frontal areas of the inferior frontal gyrus in the intact *right hemisphere* by repetitive stimulation can actually *increase* language performance such as naming, repetition, and comprehension (see Turkeltaub 2015 for a review). The explanation for these findings is that increased activity of the right hemisphere interferes with the recovery of the left lesioned side of the brain. Reducing the right hemisphere's activity 'rights' the asymmetric balance between the neural activity of the two hemispheres of the brain. The balance here is an asymmetric one; the left hemisphere is the dominant hemisphere and it must still dominate for optimal recovery.

So we now have seen some evidence suggesting the right hemisphere does contribute to language recovery in aphasia and other evidence suggesting that the right hemisphere actually impedes recovery after left brain injury. What appear to be contradictory results regarding the role of the right hemisphere in recovery may be reconcilable once taking into account more details of the available evidence. Lesion site and size and the functional role different neural areas play in language are important variables influencing the role of the right hemisphere in language recovery. Right hemisphere involvement seems greater in recovery of auditory language comprehension than in recovery of speech production, and the larger the lesion, the more likely the engagement of homologous (same location) areas of the right hemisphere (Hartwigsen and Saur 2019).

Whatever the role of the right hemisphere in language recovery, it is clear that it does not do a wholesale take-over of the language functions of the damaged left hemisphere. It is not clear whether the lack of full language recovery by the right hemisphere reflects the fact that in adults the right hemisphere has a limited language capacity and so is unable to attain the same level of linguistic function as the left hemisphere or whether it is still competing with and inhibited by the now damaged dominant left hemisphere.

One thing we do know is that time is a friend in the recovery of language in aphasia. Areas around the lesion (perilesional) become more active after the acute phase of the brain injury and appear to be critical for maximum recovery. This is presumably because a stroke or brain injury not only destroys brain tissue, but in the early stages it also produces edema or swelling in the area surrounding the damaged tissue. In addition, lesions not only produce local effects on the areas surrounding the lesion, but they also have long-distance effects called *diaschisis* (see Sect. 8.1 for further discussion). How can this be? Remember that neurons speak to each other as part of a network. As a result, injury in one part of the network has effects on neural activity in other parts of the network that may be distant from the site of the lesion. Time gives the edema in areas around the lesion a chance to resolve and for the longer distant neuronal connections to structurally and functionally reconnect with in the network to support recovery of some of the lost function caused by the original lesion.

This all sounds very depressing. But there is still hope. One critical area we have not considered is the role that speech/language therapy may play on neural reorganization. As we have just discussed, there is clearly some brain plasticity for language even in the adult. It is beyond the scope of this book to review the large field of speech pathology, the many different approaches to speech/language therapy, and recent efforts to track neural changes that occur as a result of therapeutic interventions (see Chap. 8 for further discussion). Suffice to say, there is evidence that depending on the therapy program, changes occur in neural activity in one or both of the hemispheres of the brain. Therapy either alone or in conjunction with brain stimulation or pharmaceutical procedures may ultimately contribute to the 'reactivation' of the left hemisphere, access the resources of the right hemisphere, or enhance a partnership between the two sides of the brain so that they may work in concert rather than competing with each other.

What we have seen so far is that brain plasticity does play a role in recovery of language in aphasia. We will now explore how the brain can adapt when it is deprived of an input stream from the outside world – auditory input in the case of the deaf and visual input in the case of the blind. Critically, there is no injury to the associated cortical areas of the brain. Instead, crucial input is lacking. Does the dictum, 'if you don't use it, you lose it' characterize the neural and associated effects of this deprivation on language processing or can neural plasticity compensate for this information loss? Are the areas not receiving input 'repurposed' or do they simply languish, no longer playing any functional role?

## 7.2    Speaking by Hand and Listening by Eye: Sign Language

Sign language is an alternate language used by the deaf. The eyes serve as the input channel, bypassing the ears and the auditory system, and the hands serve as the output channel, bypassing the mouth and the articulatory system. For those who do not know about sign language, it is worth briefly reviewing some of its basic characteristics.

To start off, sign language is a real language. Just as there are different languages of the world which are mutually unintelligible, there are different sign languages. For example, there is a sign language in the United States and parts of Canada (American Sign Language, ASL), a different one in Britain, in France, in China, and in Brazil to name just a few. All sign languages share the properties of spoken language. They have a set of 'speech sounds' made by the hands which, just like the sounds of language, are comprised of smaller components or features, and they have words, grammatical markers, and rules for combining words into sentences.

There are a number of important properties of sounds and words in sign. First, the 'sound' properties of the signs use three basic parameters: location, handshape, and movement of the hands. Examples of how these parameters are used in ASL are shown in Figs. 7.2, 7.3, and 7.4. Compare the signs for 'summer' and 'dry' shown in Fig. 7.2. The handshape of the two signs is the same, but the *location* of the sign is different; 'summer' is produced near the forehead and 'dry' is produced near the chin. Similarly, different handshapes and different hand movements may be used to distinguish words in American Sign Language (ASL). As Fig. 7.3 shows, both the sign for 'water' and the sign for 'vinegar' are made near the mouth, but the *handshape* is different. In contrast, the handshape and location of the signs for 'paper' and 'school' are similar, but the *movement* of the hands is different.

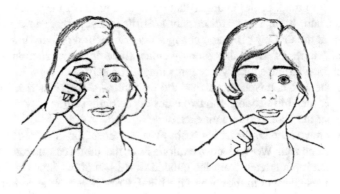

**Fig. 7.2** Examples of signs distinguished by *location*. The left panel is the sign for 'summer' and the right panel is the sign for 'dry'

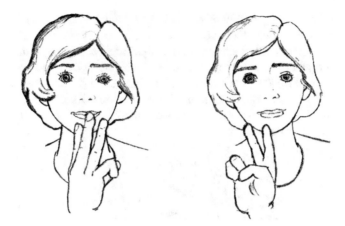

**Fig. 7.3** Examples of signs distinguished by *handshape*. The left panel is the sign for 'water' and the right panel is the sign for 'vinegar'

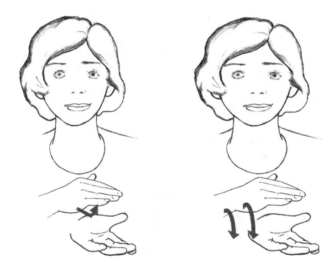

**Fig. 7.4** Examples of signs distinguished by movement. The left panel is the sign for 'paper' and the right panel is the sign for 'school'

These properties of signs are similar to the features of speech sounds we described in Chap. 3 to differentiate the sounds of language. Just as we saw in spoken language, a single feature distinguishes the word 'pear' from the word 'bear', so too signed words, as those shown in Figs. 7.2, 7.3, and 7.4, may also be distinguished from each other by a single feature. And similar to the 'slips of the tongue' and 'slips of the ear' in speech production and speech perception that we described in Chap. 3, signs distinguished by a single feature are more likely to occur in 'slips of the hand' or 'slips of the eye' than are signs distinguished by more than one feature.

A second important property of words in sign language is that the relationship between what a sign looks like and the meaning it conveys is typically *arbitrary*. What do we mean by that? It is possible that signs are *iconic*; that is, they look like what they represent or what they mean. For example, the sign for 'house' looks a bit like a house (see Fig. 7.5). For most signs in sign language, however, you cannot figure out what the meaning of the sign is by what it looks like. Let's try it. Look at Fig. 7.6, do you know from the sign what word it is? I don't think so. The sign you are looking at is the word for 'bird'.

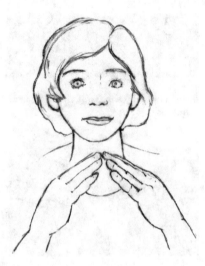

**Fig. 7.5** Example of an 'iconic' sign. The sign for the word 'house'

**Fig. 7.6** Example of a non-iconic sign. The sign for the word 'bird'

Unfortunately, deaf signers have strokes resulting in aphasia. And it is as devastating for them and their families as it is for hearing individuals. If we can distance ourselves from the human toll of aphasia, we have a unique opportunity to see whether the same neural areas are recruited for language in ASL as for spoken English, despite the fact that there are different input and output channels between the two languages.

Let's think this through. We know that different modalities are represented in different areas of the brain. Language input in sign language recruits occipital areas, not auditory areas which are recruited in spoken language, and language output recruits motor areas that represent the hands, not those that represent the mouth. Could these differences affect the representation of language in the brain?

Despite these differences, there are striking similarities between aphasia in the deaf and in the hearing. Critically, similar to language in the hearing, aphasia in the deaf is left-lateralized and typically occurs with left brain injury, not right brain injury (Hickok and Bellugi 2011). And similar to what we saw looking at the different aphasia syndromes, different areas of brain injury have different language consequences. Just like those with aphasia who hear, deaf signers with posterior lesions sign fluently but have difficulties in sign comprehension. In contrast, deaf signers with anterior lesions have relatively good sign comprehension, but are nonfluent in their production of signs, often distorting their sign articulation.

The properties of aphasia are also similar when comparing language in the deaf and language in the hearing. Recall that more substitution errors occur in spoken language between similar sounds, often resulting in two different words such as 'pear' and 'bear'. The same thing happens in sign. Here, the aphasic signer may use the wrong location, the wrong handshape, or the wrong hand movement resulting in the production of an incorrect word. For example, returning to Figs. 7.2, 7.3 and 7.4, a deaf signer might say 'summer' instead of 'dry', 'water' instead of 'vinegar', or 'paper' instead of 'school'.

Another type of sign substitution that is similar to what we described in Chap. 4 is the substitution of a word that is semantically related to the intended word. For example, instead of signing 'mother', the deaf signer might produce the sign for 'father', just as a hearing person with aphasia would.

What this tells us is the brain does not care about whether the input is by eye or by ear or whether the output is my hand or by mouth. Perhaps more crucially, the neural architecture for language is similar in the deaf and in the hearing indicating that it is the *function* not the *form* of language which dictates how it is organized in the brain.

Consistent with these findings are additional parallels between signed and spoken language. In Chap. 6, we looked at the effects of damage to the right hemisphere on language and saw that there is a functional division of labor between the left and the right hemispheres. The left hemisphere is the 'linguist', responsible for the 'pieces' of language, whereas the right hemisphere is responsible for the use of language in its social context. The same is true for sign language. Right hemisphere damage in signers results in deficits in using language in discourse. For example, there may be difficulty in staying on course with a narrative resulting in the introduction of extraneous material.

There may also be difficulties in the use of reference which is an important grammatical tool in providing coherence in discourse. Reference allows the speaker to refer back to someone or something mentioned earlier and thus provides continuity to the narrative. Let's be more concrete by constructing a short story. The first sentence of the story is 'The boy saw the girl walking down the street'. I now want to say more about the two characters in my story. Here is one possible version:

> 'The boy saw the girl walking down the street. The boy said hello and the girl smiled. The boy continued to walk down the street. The girl stopped and watched the boy walk away from the girl.'

I assume you find this version awkward and definitely not smooth in its exposition. It would be more natural to replace 'boy' and 'girl', once introduced, with the pronouns 'he' and 'she'. Reading this same narrative, now including pronouns, I think you can see it is more natural, flows more easily, and clearly marks who is being talked about. Here goes:

> 'The boy saw the girl walking down the street. He said hello and she smiled. He continued to walk down the street. She stopped and watched him walk away from her.'

In sign, reference in discourse is marked by identifying the person and 'putting' that person in a signing 'space location'. Then, when referring back to the person as the discourse continues (much as we described in the use of pronouns in our short narrative above), the signer goes back to that space without needing to resign the name of the person or who that person is. This is because the particular person is designated in this signing space. Right hemisphere damaged signers can appropriately use space to refer back to a person. However, in discourse, rather than using signing space to refer back to a particular individual, they repeat the referent. Thus, their discourse is more like that of the first story presented above than that of the second story.

We have then a parallel between signed and spoken language. Damage to the left hemisphere damages the 'linguistic' properties of language and spares its social properties, whereas damage to the right hemisphere has the opposite effect, it spares the linguistic properties of language and impairs its social and discourse properties. The spared linguistic ability of the right hemisphere is all the more surprising given a function of the right hemisphere that we have not talked about. The right hemisphere is specialized for non-linguistic visuospatial functions. Right hemisphere damaged subjects often have difficulties navigating the world; they get lost, even in an environment they are familiar with. When asked to place the location of objects in a room, they may be unable to do so. They also typically have a left neglect manifested by difficulties in processing the left-side of space. Thus, if asked to draw a flower or a clock, they may omit the left side of the flower or the left side of the clock.

Given that sign is a visuospatial language using location, motion, and shape to signify different words, one might expect that right hemisphere damage in deaf signers would have a devastating effect on the use of sign language. But it does not. As we indicated earlier, right hemisphere patients do not have a deficit in producing signed words. But they do have deficits in using space to navigate the environment. How can this be? The explanation is straightforward. Here again, as we described in Chap. 5, *function trumps form*. Space used linguistically is a left hemisphere function; space used in discourse or to navigate the world is a right hemisphere function.

It appears as though sign and spoken language use similar neural resources. But there is one critical difference. As we discussed earlier, auditory areas in the temporal lobe are deprived of input in the deaf. These areas play a crucial role in the auditory processing of language – the auditory signal is mapped to the sounds of language and ultimately maps on to words and their meanings, engaging a broad neural network involving the temporal, parietal, and frontal lobes. What happens if this first input channel to the language system can no longer receive input?

It turns out that these auditory areas are not dormant in congenital deaf signers; they are actively engaged in – guess what- visual tasks. They are recruited in processing static figures and tracking motion. Thus, neural areas functionally specialized for audition are now 'repurposed' for another modality, in this case vision. This does not mean that these auditory areas literally take over the functions of the visual system. No, the visual system in deaf signers functions as it does for you and for me. The deprived auditory areas show *cross-modality* plasticity, taking on functions that were originally auditory in nature to ones that are visual in nature. And these functions extend to the processing of sign language. Recall that sign language has all of the same properties of spoken language – it has a sound structure, words, sentences, and meaning representations. Functional neuroimaging studies show that the 'deprived' auditory areas in signers are involved in processing sign language and connect to the same networks recruited in processing words, their meanings, and their integration into sentences in spoken language.

## 7.3   Reading by Touch: Language and the Brain in the Blind

When sighted people read, visual symbols correspond to the sounds of speech and ultimately to meaning. There is a broad range of neural areas involved in this process. The visual pathway from the eyes reaches the occipital cortex of the brain (see Fig. 6.2), and from there activates cortical areas in the occipital-temporal, parietal, and frontal lobes. For those who are born blind and thus are *congenitally blind*, the visual areas are deprived of input isolating the remaining portions of the 'reading pathway' used by sighted individuals. This does not mean that the blind cannot read. All that is needed is an input system that can convert a symbol into a sound representation of language.

Think about when you were a child. Did you ever play, as I did, the game where someone outlines the shapes of letters on your back and your task is to identify the letters and/or the spelled word? Here, sensory information from the touch on your back is used to translate the shape of a letter to its representation. Braille, the reading system for the blind, also uses the tactile (touch) modality as the entry point for reading. In this case, tactile information from the hands replaces visual information from the eyes as the input channel. Braille, invented by Louis Braille who was

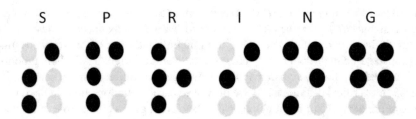

**Fig. 7.7** Braille letters associated with each letter of the word 'spring'. The dark circles correspond to the combinations of raised dots used in Braille to signify individual letters

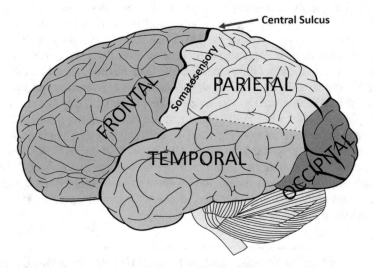

**Fig. 7.8** Location of the somatosensory cortex in the left hemisphere. The somatosensory cortex is specialized for processing sensory information such as touch

blinded in both eyes as a young boy, uses a system of raised dots with specific patterns of dots corresponding to a particular letter of the alphabet (see Fig. 7.7). The Figure shows the Braille letters associated with the word 'spring'. Braille is read by the blind and visually impaired by moving the tips of the fingers of both hands from left to right across the raised dots (signified by the filled circles) on the page.

If you were to make a hypothesis, what area of the brain do you think would be the entry point for Braille? A reasonable hypothesis would be the area of the brain specialized for touch, in this case, the somatosensory cortex, located in the parietal lobe just posterior to (behind) the central sulcus (see Fig. 7.8). Surprise! That is <u>not</u> what the neuroimaging findings show. Experiments looking at areas of the brain that are activated when blind individuals are reading Braille did not show activation in the somatosensory cortex. Rather, neural activity was shown in the visual cortex, the very area that receives no visual input in the blind.

Well, you say, perhaps the brain of those who are blind is completely rewired. Maybe all tactile information is processed in the occipital cortex, not just reading Braille. Nope – not the case. The somatosensory cortex of the blind is activated, as it is for sighted individuals when blind individuals are asked to make a judgment based on touch, such as deciding whether an object is rough or smooth, or its dimensions are straight or curved. Thus, the visual cortex of the blind appears to be 're-purposed' for tactile reading. More generally, it shows that the brain is indeed plastic. Neural areas thought to be hard-wired and functionally specialized for one modality, in this case vision, have taken on a 'function' of a different modality, touch.

There is more to this story. Visual areas appear to be recruited in congenitally blind individuals not only when reading Braille, but more generally in processing language. For example, these areas are activated when congenitally blind individuals hear a noun, e.g. 'dog', and are asked to produce a verb related to it, such as 'bark' (Burton et al., 2002), or when they are asked to comprehend sentences that they hear (Bedny et al., 2011). What this means is that the brain is 'using' the visually deprived areas in the congenitally blind. Not only are frontal, temporal, and parietal areas recruited in language processing, but visually deprived areas contribute to language as well. But there are also limits. Just as we discussed in the introduction to this chapter, the extent of plasticity of the visual areas for assuming language functions is dependent on the age at which someone becomes blind. The greatest plasticity occurs for those who are born blind, less for those who are blinded as children, and minimally, if present at all, in adults (16 years and older).

## References

Bedny, M., Pascual-Leone, A., Dodell-Feder, D., Fedorenko, E., and Saxe, R. (2011). Language processing in the occipital cortex of congenitally blind adults. *Proceedings of the National Academy of Sciences, 108(11)*, 4429–4434.

Hartwigsen, G., and Saur, D. (2019). Neuroimaging of stroke recovery from aphasia–Insights into plasticity of the human language network. NeuroImage, 190, 14–31.

Hickok, G., and Bellugi, U. (2011). Neural organization of language: Clues from sign language aphasia. J. Guendouzi, F. Loncke, and M. J. Williams, (Eds.). The Handbook of Psycholinguistic and Cognitive Processes. New York: Taylor and Francis, pp. 687–708.

Posner, M.I. and Raichle, M.E. (1994). Images of Mind. New York: Scientific American Library.

Turkeltaub, P. E. (2015). Brain stimulation and the role of the right hemisphere in aphasia recovery. *Current Neurology and Neuroscience Reports, 15(11)*, 1–9.

## Readings of Interest

Amedi, A., Merabet, L. B., Bermpohl, F., and Pascual-Leone, A. (2005). The occipital cortex in the blind: Lessons about plasticity and vision. Current Directions in Psychological Science, 14(6), 306–311.

Burton, H., A. Z. Snyder, J. B. Diamond, and M. E. Raichle. "Adaptive changes in early and late blind: a FMRI study of verb generation to heard nouns." Journal of neurophysiology (2002).

Dehaene, S. (2009). Reading in the Brain. New York: Viking Press.

DeMarco, A. T., and Turkeltaub, P. E. (2018). Functional anomaly mapping reveals local and distant dysfunction caused by brain lesions. BioRxiv, 464248.

Hickok, G., Bellugi, U., and Klima, E. S. (2001). Sign language in the brain. Scientific American, 284(6), 58–65.

Kiran, S., and Thompson, C. K. (2019). Neuroplasticity of language networks in aphasia: advances, updates, and future challenges. *Frontiers in Neurology*, *10*, 295.

# Beyond Aphasia: What More Do We Know

**8**

The goal of this book has been to understand how brain injury affects language in aphasia. This has provided a unique perspective not only on aphasia but also more broadly on the neural basis of language. Indeed, the study of aphasia has and continues to provide the foundation for what we know about language and the brain. However, there are approaches that use different technologies, methods of testing, and participants other than those we have considered thus far. As such, they may come up with findings that reaffirm, extend, or challenge those that we have drawn about the nature of language impairments in aphasia. In this chapter, we will examine a number of these approaches, always with an eye towards informing and enriching what we have learned about aphasia.

## 8.1 Distributed Neural Systems in Language

Research using advanced technologies that allow for mapping language and other cognitive functions in the intact as well as the lesioned brain has flourished in the last 50 years (for a historical review see Posner and Raichle 1994). One set of methods uses functional neuroimaging which allows for millimeter mapping of brain activity while subjects are performing a task. Those areas of the brain that are activated are considered to play a functional role in the subjects' performance.

One such method, functional magnetic resonance imaging (fMRI), is a non-invasive technique that measures changes in blood flow and oxygen level uptake. Greater blood flow and oxygenation reflect the increased neural resources dedicated to specific brain areas during language processing. There is a large fMRI literature investigating the neural areas involved in processing those aspects of language that we have examined in preceding chapters looking at aphasia. Just as we did, these neuroimaging studies have examined the perception of the sounds of speech by looking at how listeners perceive differences between sounds and the features underlying them. They have studied the processing of words by examining how the activation of one word may influence the activation of another word that shares

© Springer Nature Switzerland AG 2022
S. E. Blumstein, *When Words Betray Us*,
https://doi.org/10.1007/978-3-030-95848-0_8

sound or meaning attributes. And they have explored the neural areas involved in processing sentences by looking at how the brain responds to sentences which vary in syntactic complexity. The majority of these studies map neural activity in individuals without brain injury providing a means for us to compare similarities and differences in the neural areas recruited in the uninjured brain compared to the injured brain in aphasia.

Without going into the fine details, the neural systems recruited in processing speech, words, meaning, and/or syntax using fMRI or other neuroimaging methods are similar to those areas we have identified in our study of aphasia. Results of these studies typically show that the processing of the different components of language are not narrowly localized in a focal area of the brain. Rather, each of them is broadly represented activating multiple neural areas in the left hemisphere including frontal, temporal, and also parietal structures. Indeed, neuroimaging studies that look at how neural areas interact with each other show that during the performance of tasks involving speech, words, meaning, or syntax, activation in one region of the brain typically influences activity in another region indicating that they are functionally connected.

These neural systems are not only functionally connected to each other but they are also neurally connected. What does this mean? Remember our description of the neural architecture of the brain in Chap. 2. Neurons connect to other neurons in a network. When and how much they fire is determined by the connections they have with other neurons. Consider what would happen with neural damage to some neurons that are part of a particular network. Damage to these neurons will not only affect how they function, but it will also influence the 'behavior' of neurons that may be more distant in that network. And that is exactly what happens in the brain to the networks supporting language.

Typically, we think of aphasia as caused by a brain lesion to a particular area or areas of the brain. A lesion causes structural damage to brain tissue. It is generally assumed that areas where there is no lesion are undamaged and hence are functioning normally. However, it turns out that neuroimaging methods such as positron emission tomography (PET) which measures metabolic activity of the brain show that in addition to changes in lesioned areas, neural areas that appear to be anatomically normal are not. They show a reduction in their metabolic activity as measured by their usage of glucose and oxygen, the lifeblood of neurons. Thus, lesions produce both local effects which reflect the structural damage caused by a stroke or other injury, and long-distance effects (*diaschisis,* see sect. 7.1) which reflect a reduction in the ability of the connected neural tissue to function normally (DeMarco and Turkeltaub 2018).

Returning to aphasia, using the lesion method, we have described the occurrence of Broca's aphasia as resulting from frontal lesions, and the occurrence of Wernicke's aphasia as resulting from temporal lesions (see 2.4.1). However, based on the similarity of patterns of errors between these two syndromes discussed in Chaps. 3, 4, and 5, we concluded that the components of language including speech, words, meaning, and sentences are not narrowly localized but rather, each component is comprised of a distributed system involving both frontal, temporal, and often parietal lobe regions. The similarity of patterns of errors between Broca's and Wernicke's aphasia presumably reflects the connectivity between neural regions. That is, frontal, temporal, and parietal areas that may underlie a particular component of language are not separated either neurally or functionally because together they form a network. Lesions in frontal regions will have long-distance effects on temporal and parietal regions even if those areas are not lesioned, and lesions in temporal regions will have long-distance effects on frontal and parietal regions even if those areas are not lesioned. As a consequence, damage in one part of the neural system affects the activity and hence performance in another part of the system.

## 8.2    Functional Differences within Neural Areas Processing Language

Neuroimaging studies have done more than reaffirm the findings from aphasia. They extend and deepen what we know about language and the brain. One of the questions which these studies have addressed that goes beyond what we have learned thus far is whether there are functionally specialized areas associated with each component of language localized within the broad neural networks we have identified. Recall that lesions producing aphasia tend to be large and hence may encompass areas not functionally related to a particular language component. In contrast, fMRI shows activation patterns in millimeters allowing for the identification of 'functional' areas involved in the processing of speech, words, meaning, and syntax much more precisely than can lesion analyses.

Localization information using fMRI has shown that particular areas within the frontal and temporal lobes are recruited in processing the different components of language (see Figs. 8.1 and 8.2). Within the temporal lobe, the superior temporal cortex, and within the frontal lobe, the pars opercularis, are recruited in speech processing. Similarly greater detail is provided in characterizing the neural areas involved in the processing of words and sentences. Here, the superior temporal gyrus and the middle temporal gyrus within the temporal lobe, the pars triangularis within the frontal lobe, and areas within the parietal lobe including the supramarginal gyrus and the angular gyrus are activated in word processing. Syntax and sentence processing recruit a broad set of areas including the anterior temporal lobe, the middle temporal gyrus, and the pars triangularis. Language areas activated in the processing of meaning, whether involving words or sentences, activate temporal areas including the middle temporal gyrus and anterior temporal lobe, parietal areas including the angular gyrus, and the frontal lobe including the pars triangularis and the prefrontal cortex (Binder et al. 2009).

So we know now that there are more circumscribed areas within the temporal, frontal, and parietal temporal lobes that are involved in processing the components of language. What we don't know is whether there are functional differences within these more localized areas. In our discussion of the components of language in Chaps. 3, 4, and 5, we talked about each neural area as a unitary whole and did not consider whether there are functional divisions within each of these systems. It

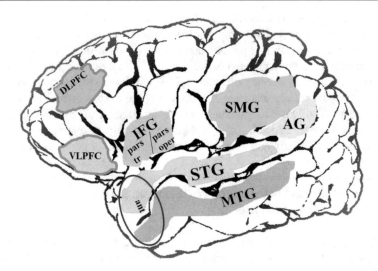

**Fig. 8.1** Neural areas involved in processing the components of language. These areas include the MTG (middle temporal gyrus), the STG (superior temporal gyrus) and ant STG (anterior STG) in the temporal lobe; the IFG (inferior frontal gyrus) encompassing the pars tr (pars triangularis) and pars oper (pars opercularis) in the frontal lobe; the SMG (supramarginal gyrus) and AG (angular gyrus) in the parietal lobe. The VLPFC (the ventrolateral prefrontal cortex) and DLPFC (the dorsolateral prefrontal cortex) in the frontal lobe are part of networks recruited in cognitive functions that can affect language processing

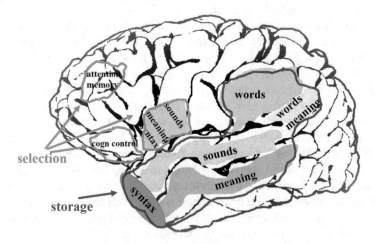

**Fig. 8.2** Functions associated with particular brain areas. The neural areas associated with these functions are shown in Fig. 8.1

turns out that there is more to the story and the information from aphasia presented earlier was incomplete. Indeed, there are multiple stages of processing within each component of language; information whether relating to sounds, words, meanings, or syntax needs to be stored, it needs to be accessed, and it needs to be selected.

An example might be useful here. Consider what we learned in Chap. 3 when we described the speech component. When you listen to speech, you hear an acoustic input which varies in frequency, amplitude, and time. This is the first stage of processing the sounds of speech. The acoustic signal is then mapped on to phonetic features associated with a particular sound. Not only is the sound activated but other sounds that share features with it are also partially activated. That is another stage of processing. But the job of perceiving speech is not done. You, the listener, must select the sound from the possible sounds which share features with it and you must ultimately make a decision about it or act upon this information. As you can see, perceiving speech involves multiple stages of processing. The question which researchers are asking is whether these separate stages recruit different neural areas within the speech component. The data suggest there is a division of labor within the neural areas representing the speech component. One stage of processing involves the acoustic analysis of speech, another is the mapping of the acoustics to features, and still another is the selection of the sound from the set of sounds that are activated. As well, there are functional subdivisions within each of the other components of language.

Functional subdivisions do not just pertain to the stages of processing language properties. After all, language is a part of a larger complex cognitive system. Language communication requires that we be able to attend to a conversation, remember what was said before, have a goal or plan as we try to make a point, and integrate a linguistic message with the context of an interchange with someone or with an event. We need to select what we want to say or select the appropriate word or meaning that we hear from a range of alternatives. To do this, we draw on other cognitive systems called executive functions which include attention, memory, and cognitive control.

Evidence from functional neuroimaging has provided a level of detail about the multiple stages contributing to our processing of sounds, words, and sentences that was impossible in our study of aphasia. It is not our purpose to provide a detailed description of these findings, but instead to highlight some important functional differences that have emerged from this research.

There appears to be a division of labor between temporal and frontal areas. This division of labor reflects the *storage* and *activation* of possible sound and word candidates, on the one hand, and the ultimate *selection* of and *decision* about the identity of a particular candidate, on the other (see Fig. 8.2). FMRI results suggest that storage and activation stages of processing recruit temporal areas, with the superior temporal gyrus activated in the processing of sounds and the middle temporal gyrus and anterior temporal lobe activated in the processing of meaning and sentences. In contrast, selection and decision stages of processing for sounds, words, meaning, and sentences occur in frontal areas, particularly in the inferior frontal gyrus and in the dorsolateral prefrontal cortex (DPFLC) and the ventrolateral prefrontal cortex (VPFLC).

The principles that underlie storage and activation stages, on the one hand, and selection and decision stages, on the other, are consistent with the functional architecture of language we have described in earlier chapters. Recall that the activation of a speech sound partially activates other sounds which share feature properties. Similarly, the activation of a word partially activates other words which share sound, meaning, or grammatical role. These different stages of processing recruit different portions of the neural areas associated with each of the components of language. In each, temporal areas are involved in the storage and activation of a particular target and along with parietal areas the co-activation of a set of 'related' sounds and words. In both speaking and listening, frontal areas are involved in the selection of the particular target from these competing alternatives based on its level of activation relative to the other competitors.

In Chap. 2, we indicated that the lesions giving rise to Wernicke's aphasia included the temporal lobe and could extend to the parietal lobe. However, we did not discuss any potential functional role that the parietal lobe might play. FMRI studies suggest that several areas within the parietal lobe, the supramarginal gyrus and the angular gyrus in particular, are involved in processing the sounds of words and the meanings of words (see Figs. 8.1 and 8.2).

The supramarginal gyrus has shown sensitivity to words as a function of their sound similarity – a word with a lot of sound competitors shows increased neural activation compared to a word that has few sound competitors. To illustrate this, let's compare the word 'bear' with the word 'elephant'. There are many words that share sounds and compete with 'bear' including 'pear', 'tear'. 'dare', 'care', 'share', 'bail', 'bag', 'bass', to name a few. In this sense, just like the Mr. Rogers TV show of the 60s and beyond, these words are neighbors. Now think about 'elephant'. How many words can you come up with that are similar in sound to it? I cannot come up with any. Even if you do better than I and can think of a few, the bottom line is that there are very few words that compete with the word 'elephant'. The greater activation shown in fMRI studies for words that have a lot of competitors tells us that it takes more neural resources to select a word when there are lots of competitor words activated compared to when there are only a few competitors.

As to the angular gyrus, it appears to play a role in the retrieval of meanings of individual words. It also appears to play a role in integrating the meaning of words within a sentence and across sentences. Thus, this area is implicated in comprehension of sentences as well as comprehension of discourse or conversations between individuals.

## 8.3   Representations for the Sounds of Language: Reaffirming What We Know

In Chap. 3, we described the architecture of the sounds of speech for production and perception. There we indicated that speech sounds are broken down into smaller pieces or features, and that these features are defined in terms of their acoustic-articulatory patterns: acoustic attributes used in perception and articulatory commands used in production. As we saw, speech perception and speech production deficits in aphasia reflect this feature architecture, with greater numbers of errors occurring for sounds distinguished by one feature compared to several features.

Additionally, lesions to the superior temporal gyrus result in speech perception deficits and auditory comprehension impairments in Wernicke's aphasia.

Recent electrophysiological evidence from electrocorticography (ECoG) supports this feature-based architecture. ECoG measures electrical activity on the cerebral cortex. It is used to identify the site of activity in individuals who have intractable seizures with the goal of surgically excising the damaged tissue. While doing this lifesaving procedure, it is also possible to examine how the brain responds to different types of stimuli. In one study (Mesgarani et al. 2014), a grid of electrodes was placed on the superior temporal lobe in seizure patients to record electrical activity on the exposed portion of the brain. These patients were presented with hundreds of sentences (500) spoken by a large group of speakers (400). Results showed that different electrodes responded to individual features of speech similar to the ones we described in Chap. 3. For example, some electrodes 'fired' to stop consonants, 'p t k b d g', others fired to fricative consonants, 'f s sh z', others to nasal consonants, 'm n ng', and still others to different features for vowels.

These findings are remarkable. Think about it. We know that the way different speakers produce sounds and the contexts in which sounds occur affect the acoustic realization of speech. For example, the acoustics of my saying 'He put the butter on the table', and your saying the same sentence is not the same. We have different vocal tract sizes (I am a small-sized female), we likely have different regional accents (I am originally from New York), and the production of the sounds in the sentence varies depending on where it appears in a word and what vowels precede or follow it. In this example, the sound 't' varies and is acoustically different when it is in the beginning of a word, 'table', end of a word, 'put', or in the middle of a word, 'butter'. And yet, listeners will say we produced the same sentence and agree that the same sound 't' appears in the three words 'table', 'put' and 'butter'. Now consider the number of different people we interact with and the thousands of different sentences and words that we produce. Given this tremendous variability in the acoustic realization of speech, it might appear that perceiving speech as we do so easily, quickly and accurately would be difficult, if not impossible. But it is not. The ECoG findings show that the brain is able to strip away the variability in speech and extract common patterns across different speakers and contexts, and those patterns correspond to features.

## 8.4  The Right Hemisphere and Language: Still a Puzzle

As described earlier, neuroimaging studies often find additional areas activated during language tasks that are not typically shown in lesion studies of aphasia. Perhaps the most consistent finding is activation in non-brain damaged individuals in the right hemisphere across tasks that tap the components of language – sounds, words, meaning, and syntax. This activation tends to be weaker than that in the left hemisphere, but it typically *mirrors* the same areas as those activated on the left.

The question is does this right hemisphere activation shown in fMRI studies indicate that the right hemisphere is playing a functional role in language for you and I. Unfortunately, just as we were unable to answer this question in Chap. 6, it is unclear whether the fMRI results indicate that the right hemisphere is playing an active role in language processing along with the left hemisphere in the uninjured adult brain. The reduced activation might suggest that the right hemisphere has some capacity for processing language but it is playing a weaker role than that of the left hemisphere. Or it could mean that its capacity is so weak as to be for all intents and purposes non-functional.

## 8.4.1  There Is a Method to Our Madness

Where do we go from here? Fortunately, there are other methods that may provide some answers. Similar to fMRI, these methods can measure neural responses in healthy adults, providing a picture of the right hemisphere's capacity for language absent brain injury. Event-related potential (ERP) is one such method, providing a non-invasive electrophysiological measure of electrical activity of the brain. Electrodes, placed on the scalp, measure electrical activity time-locked to a particular stimulus. In this way, it is possible to examine how the brain responds to a particular event such as the presentation of a word or a sentence.

ERPs show sensitivity to the meanings of words. Different responses occur to a word like 'fruit' when it is preceded by a semantically related word such as 'pear' compared to when it is preceded by a semantically unrelated word such as 'nose'. ERPs also show sensitivity to the context in which a word occurs in a sentence. Different patterns of response occur when a word fits semantically with the context of a sentence compared to when it does not. For example, 'butter' is semantically congruent with the preceding context in 'he went to the store to buy bread and butter', but it is semantically incongruent with the preceding context in 'he went to the store to buy shoes and butter'. Such findings suggest that this technique can be used to examine the semantic processing of the two hemispheres. As you may recall from Sect. 6.2, lesion studies showed coarse semantic coding in the right hemisphere. That is, in contrast to the left hemisphere, the right hemisphere failed to show sensitivity to small differences in the semantic features distinguishing words. Does this occur in the healthy brain as well?

Let's say you were given the sentence, 'Every morning John made a glass of freshly squeezed juice. He keeps his refrigerator stocked with ___'. What word would best end the sentence? I assume you would say, 'oranges'. Although that is the most likely response, there are other possible responses. Less likely, you could squeeze another fruit like 'apples'. Or even a little less likely, but still possible, you could squeeze a vegetable like 'carrots'. As you can see, the words vary in the degree to which they are likely to fit into the preceding context.

ERP responses to these words presented to either the right or left hemisphere preceded by the same sentence context showed different patterns of response (Kutas and Federmeier 2000). The left hemisphere produced a gradient response to the words as a function of how well they fit into the context, showing sensitivity to the fine-grained differences in the semantics of the word to fit into the preceding context. That is, given the context 'Every morning John made a glass of freshly squeezed juice. He keeps his refrigerator stocked with …', the left hemisphere showed the greatest sensitivity to 'oranges' given this preceding context, a little less sensitivity to 'apples', and still less to 'carrots'. In contrast, the right hemisphere showed the greatest sensitivity to 'oranges', the word that was most likely to fit in the context, but treated the two less likely words the same. In other words, the right hemisphere failed to show a gradient response, but rather displayed a dampened sensitivity to the semantic differences between words that were 'more distant' from the word that fit best in the context.

These results tell us that the right hemisphere in the healthy adult brain has the *capacity* to process the meanings of words. Just as we saw with individuals with lesions to the right hemisphere described in Sect. 6.2, the right hemisphere does not differentiate distinctions within the semantic network to the same extent as the left hemisphere, leaving it unable to capture the richness of the meanings inherent in the words of language that are captured by the left hemisphere. In this sense, the right hemisphere has a limited semantic capacity. Nonetheless, what the ERP findings cannot tell us is whether the right hemisphere in the adult brain is passively processing meaning or is playing an active functional role in concert with the left hemisphere.

There is one method, transmagnetic stimulation (TMS) that may be able to speak to whether the right hemisphere is actively involved in processing language in the healthy adult brain. TMS is a non-invasive electrophysiological method that can produce temporary 'virtual' lesions by sending electromagnetic pulses through the skull and disrupting normal brain activity in targeted brain areas. Of particular significance, this procedure may be used to examine potential disruptive effects in the healthy (non-brain-injured) adult brain, providing an important comparison to the effects of lesions caused by brain injury. Just as we have learned that lesions in aphasia have both local and long-distance effects, so too do 'virtual' lesions produced by TMS, allowing for mapping out the neural networks contributing to language processing.

Will virtual lesions to specific brain areas in the right hemisphere affect language performance, and, if so, will the magnitude of the 'deficit' be equal to that of virtual lesions created in the left hemisphere? Will the patterns of deficit be similar between the two hemispheres? And will they mirror the effects of lesions caused by brain injury? To date, there are only a handful of TMS studies that have compared left and right hemisphere processing of specific aspects of language. We will have to await further research to give us the answers we seek. However, this method holds the promise of providing unique insights into the role of the right hemisphere in processing language in the healthy adult brain.

## 8.5    It Depends

Without doubt, we have learned new information about language and the brain by looking at other approaches. This is true for any science. One methodology is not enough to answer any scientific question. A question you may now be asking is whether these new approaches supplant the study of aphasia. In other words, why study aphasia if we can now get greater detail about the neural systems underlying language? The answer is: it depends. It depends upon what questions we are asking and what we want to learn. It also depends upon whether aphasia can continue to contribute data and information that the new approaches cannot.

If the goal of a researcher is to understand the details of how the brain instantiates language, then the study of aphasia has provided the first step. New approaches clearly can provide details and insights that may not be derived from focusing solely on aphasia. However, if the goal is to understand the effects of brain injury on language, on the one hand, and to facilitate language recovery, on the other, then aphasia has to be center stage.

There is always more to learn from the study of aphasia. Keep in mind that the varying clinical picture shown by those with brain injury provided researchers interested in understanding the neural bases of language with the first clues on the role of different neural areas involved in speaking, understanding, and communicating using language more generally. And such clues continue to be provided by the study of individuals who sustain brain injury by applying the new techniques to those with aphasia.

Additionally, brain injury may hold surprises not predicted or expected. This is apparent in individual case studies where unexpected symptoms appear after an individual has sustained brain injury. Two examples are instructive. The first stems from a case study reported by Joseph Dejerine (Dejerine 1892). His patient had a reading deficit (alexia). However, the unexpected finding was that his patient could still write, although he was unable to read what he wrote. This syndrome called alexia without agraphia is rare and is found with a lesion profile involving the left visual cortex and the splenium of the corpus callosum (Geschwind 1965). This syndrome provided unique insight into the neural basis of reading and would not have been identified without a clinical exam of this particular patient and subsequent analysis of the lesion.

The second example is the foreign accent syndrome, first noted by Pierre Marie (Marie 1907). In this syndrome, the patient speaks with what sounds like a foreign accent. The articulatory patterns differ from the speech output of those with Broca's aphasia; speech is 'off' but it is not distorted, nor is it nonfluent. Rather, the production of vowels and the rhythm and timing of speech is affected. Interestingly, studies have shown that this syndrome does not produce a true foreign accent at all, but rather is perceived by the listener as a non-native accent.

Who would have predicted that a brain injury to the speech-motor system would render a person sounding like a 'foreigner' speaking in their native language? What changes to the speech motor control system and what neural areas give rise to this unique speech pattern? More research is needed to answer these questions. However, what is clear is that only clinical examination and experimental research with brain-injured patients would have identified the speech output deficit associated with this syndrome.

# References

Binder, J. R., Desai, R. H., Graves, W. W., & Conant, L. L. (2009). Where is the semantic system? A critical review and meta-analysis of 120 functional neuroimaging studies. Cerebral cortex, 19(12), 2767–2796.

Dejerine, J.J. (1892). Contribution a l'etude anatomo-pathologique et Clinique des differentes varieties de cecite verbale. Memoire de la Societe de Biologie, 4, 61–90.

DeMarco, A. T., and Turkeltaub, P. E. (2018). Functional anomaly mapping reveals local and distant dysfunction caused by brain lesions. BioRxiv, 464248.

Geschwind, N. (1965). Disconnexion syndromes in animals and man. Brain, 88(3), 585–585.

Kutas, M., and Federmeier, K. D. (2000). Electrophysiology reveals semantic memory use in language comprehension. Trends in cognitive sciences, 4(12), 463–470.

Marie P. (1907). Presentation de malades atteints d'anarthrie par lesion de l'hemisphere gauche du cerveau. Bulletins et Memoires Societe Medicale des Hopitaux de Paris, 1, 158–160.

Mesgarani, N., Cheung, C., Johnson, K., and Chang, E.F. (2014). Phonetic feature encoding in human superior temporal gyrus. Science, 343(6174), 1006–1010.

Posner, M. I., and Raichle, M. E. (1994). Images of Mind. Scientific American. New York: Library/Scientific American Books.

# Readings of Interest

Badre, D. 2020). On Task: How Our Brain Gets Things Done. New Jersey: Princeton University Press.

Bates, E., Wilson, S. M., Saygin, A. P., Dick, F., Sereno, M. I., Knight, R. T., and Dronkers, N. F. (2003). Voxel-based lesion–symptom mapping. Nature Neuroscience, 6(5), 448–450.

Cheung, C., Hamilton, L.S., Johnson, K., and Chang, E.F. (2016). The auditory representation of speech sounds in human motor cortex. Elife, 5e, 12577.

Devlin, J. T., & Watkins, K. E. (2007). Stimulating language: insights from TMS. Brain, 130(3), 610–622.

Fridriksson, J., den Ouden, D. B., Hillis, A. E., Hickok, G., Rorden, C., Basilakos, A., … and Bonilha, L. (2018). Anatomy of aphasia revisited. Brain, 141(3), 848–862.

Mirman, D., Chen, Q., Zhang, Y., Wang, Z., Faseyitan, O. K., Coslett, H. B., and Schwartz, M. F. (2015). Neural organization of spoken language revealed by lesion-symptom mapping. Nature Communications, 6:6762.

Ryalls, J. and Miller, N. (2014) Foreign Accent Syndromes: The Stories People Have to Tell. Oxfordshire, England: Routledge

Thompson-Schill, S. L., Bedny, M., and Goldberg, R. F. (2005). The frontal lobes and the regulation of mental activity. Current Opinion in Neurobiology, 15(2), 219–224.

# A Message of Hope

**9**

In the introduction to this book, we promised to tell the story of aphasia. To do so, we examined how language breaks down as a result of brain injury and how such study provides a window into language and brain function. We took a scientific approach and showed how experimental research provides insights into the intersection of language and the brain. However, there has been one part of the story that has been largely missing. That is the human side of aphasia. What we have not considered, except in passing, is the people who have aphasia and their personal stories. As a result, although you may now have deeper insight on the effects of brain injury on language function, your understanding is divorced from the realities of aphasia – its feel, its emotion, and its impact on the lives of those touched by this disorder. You also do not have a sense of how research in the lab (the *bench*) impacts and is translated to the patient (the *bedside*). In this chapter, we hope to tell some part of this other story.

## 9.1 The Many Stories of Aphasia: Meeting Challenges Head-On

Aphasia does not discriminate – it can strike anyone. There is a long list of notable figures who have had aphasia including Charles Baudelaire, the French poet; Julie Harris, the actress; Kirk Douglas, the actor; Dick Clark, the entertainer; Michael Hayden, U.S. Air Force General; and Dwight David Eisenhower, the 34th President of the United States, among others (Profiles of notable individuals with aphasia can be found on the National Aphasia Association website; https://www.aphasia.org/). Each person with aphasia has a unique story to tell. Several vignettes are described below that give you a flavor of those stories. Each vignette not only tells a unique story but also sends a message to you, the reader, a message that can only be captured in human terms. Together they provide a window into the strength of the human spirit in responding to and coping with aphasia.

© Springer Nature Switzerland AG 2022
S. E. Blumstein, *When Words Betray Us*,
https://doi.org/10.1007/978-3-030-95848-0_9

### 9.1.1 A Program to Recovery

Patricia Neal was an accomplished actress who won a Tony Award in 1947 for her performance in the play *Another Part of the Forest* and an Academy Award in 1963 for her performance in the film *Hud*. She had a series of strokes in 1965 and was left with a hemiplegia (paralysis of her right arm and leg) and a severe aphasia marked by difficulty in speaking. Indeed, she was often unable to come up with words and would often produce neologisms such as 'skitch' and 'oblogon' as she tried to talk (Farrell 1969, pp. 38–39).

Speech therapy programs typically have scheduled meetings several times a week that may last from one-half hour to an hour. Her husband, Roald Dahl, the noted author and screenwriter, thought this was simply not enough time spent working on Patricia Neal's language impairment. He believed that she should be placed in an intensive working regimen where she would be constantly exposed to language and other 'brain exercises'. To this end, in addition to her scheduled speech therapy, Dahl engaged a cadre of close friends and neighbors to interact and work with her throughout the day. Indeed, she was engaged with others daily from 9 am to 6 pm with a 2 or 3 hours break for lunch and rest.

Farrell (1969) narrates the poignant story of her struggle and provides a window into the support network and intense program of rehabilitation created by Dahl. Such intense therapy is rare – it requires resources – financial, people, time, and a planned program. The idea has its roots in total immersion programs in second language learning. Being in an environment where the learner literally lives, eats, and breathes in the language mimics how the brain learns and responds to language. It does so in a more natural way than training exercises that happen only a few hours a week.

Dahl organized and wrote out a detailed therapy plan with instructions for the 'crew' of friends, all amateurs with no training in speech rehabilitation. The plan included a varied set of exercises and tasks that required all means of interaction and communication; they were to read books, play board games and card games, name objects and pictures, do jigsaw puzzles, put sentences together from a given a set of words, and do math problems. As you can see, the object was not only to stimulate Patricia Neal's use of language but also to stimulate her cognitive abilities more generally. She was to be engaged, and her 'teachers', as they were called, were not to let her give up. Indeed, Dahl recognized the difficulties and frustration that Patricia was facing. He instructed the teachers, 'Push, push, push for answers, make the patient sit up, try like mad, dig out something from somewhere, mangle out a bit of vitality, a laugh, an *effort*' (Farrell 1969, p. 150).

There are two major challenges that can affect the recovery of someone with aphasia and that can work against the intense regimen dictated by Dahl's program. The first has to do with the patient. Patricia Neal was painfully aware of her deficit – her inability to say what she wanted to say; her failed attempts to get at the words she wanted to express; her embarrassment at her now diminished language abilities; her fear that she was not nor would ever be the same person she had been; her impairment in using her right arm and leg. The program was not only difficult, but it was exhausting and frustrating for her.

The natural tendency for anyone with aphasia is to not want to talk or to communicate under such circumstances, much as any of us avoid doing things that are particularly hard and potentially embarrassing. It takes grit and determination to use language daily over many hours without giving up. The degree of frustration and embarrassment cannot be underestimated. I witnessed one concrete example in an interview with a person with aphasia as he recounted an interaction he had with his young son. Overall, his recovery was excellent. His speech was fluent and he understood what was being said. He had some residual difficulties, one of them being reading. One day, he was reading the newspaper and got stuck on a word. He couldn't make it out and called over to his young son for help. His son looked at the word, then his father, and read the word to him. The word was 'it'. In the interview, the aphasic said the word 'it' and then, spelled out the letters individually, 'i', 't', shaking his head in disbelief and in embarrassment that he had to ask his little boy for help in reading such a simple word.

To avoid such situations, some of those with aphasia would simply not have asked the question. Fortunately, this man and Patricia Neal had the courage, yes courage, to push themselves even in the face of failures. Both had the trust and support of family and friends. The role that caregivers play in supporting and helping persons with aphasia cannot be underestimated. This brings us to the second challenge that can influence the recovery process in aphasia.

It is painful to watch a loved one struggle to speak. A natural inclination is to try to anticipate what the patient is trying to communicate. That way, it is possible to avoid the pain of watching a loved one experience the frustration, embarrassment, and even anger that can accompany failed attempts. It requires inordinate patience to not fill in gaps as the patient struggles to speak and come up with a particular word. Indeed, it is not uncommon for there to be silent gaps lasting seconds and even minutes in a conversation with someone who has aphasia. Filling in the word or even anticipating the needs of the patient may seem like a kind and helpful thing to do. However, by doing so, the caregiver is not doing the patient any favors. Change and improvement cannot happen in silence. In his training regimen, Roald Dahl had to be sure that those who had personal connections to Pat but were not professional speech pathologists understood that they had to stay the course and insist that she talk and stay engaged. They had to hear the mistakes, watch her struggle, but encourage her to push on.

Like the song, for there to be progress, improvement, and change, it is necessary to 'pick yourself up, dust yourself off, and start all over again!' (composer Jerome Kern, lyricist Dorothy Fields). And that is exactly what Patricia Neal did. Her road to recovery was daunting, and although she never attained the level of language she had prior to the stroke, she soldiered through. Patricia Neal's lifeblood was on stage, and she returned to it in 1968, 3 years after her stroke. She played a leading role in the movie *The Subject was Roses* where she was nominated the following year for an Oscar for her performance.

## 9.1.2   Never Give Up

Gabby Giffords, a member of the United States House of Representatives from 2007 to 2012, was the victim of an assassination attempt on January 8, 2011 during an outdoor event, "Congress on Your Corner", that she was holding with her constituents. She was shot in the head and sustained serious brain injury that affected her language, her vision, and her ability to use her right side.

A glimpse of her early days and the magnitude of her language deficit can be seen in an ABC news report with Bob Woodruff (https://www.youtube.com/watch?v=rx3nfUKvrZ8). At the outset, Gabby Giffords could not say anything. A later CBS interview with Lee Cowan, March 15, 2015, held some four years after her injury, shows how much improvement she had made up to that point (https://www.youtube.com/watch?v=4_kTv2t2hlY&list=RDtiJ9X_wLSWM&index=15). In that interview, she and her husband, Mark Kelly, the astronaut and now senator from Arizona, describe the early days of her struggle with aphasia. Her aphasia was very severe – she could produce at most a single word. Frustrating for Gabby, she would often get stuck on a particular word. For some time, 'chicken' was the word. It would just pop out, even if it was not appropriate to the context. Gabby could modulate the intonation or pitch of her voice when saying the word which gave it some 'meaning'. But she was not able to 'break away' from it. Such perseverations are a common feature for some who have a non-fluent aphasia as she did.

As shown in the Cowan interview, the improvement she had made was substantial. Although she had trouble articulating her words, she was now able to speak in short phrases. She was clearly engaged in the conversation, understanding what was said. Most recently on August 19, 2020, Gabby gave a speech at the Democratic National Convention, some 9 years after her injury. There, she spoke slowly using full sentences. Her presentation was clear and forceful. For those of you who want to see the dramatic changes in her language over time, compare her speech in the ABC and CBS interviews with her complete speech at the convention which can be found on the internet. The full story of her life and recovery is described in detail in the book entitled *Gabby: A Story of Courage, Love, and Resilience* (Giffords and Kelly 2012).

Gabby Giffords' road to recovery has been long and arduous. She has had intense speech therapy that has continued over time, and she has also received consistent and committed support from her husband, two essential ingredients for a road to recovery. Support and an advocate are essential ingredients. Much of the focus of her speech therapy was on speaking – articulating sounds, words, and slowly increasing the amount of speech she could say until she could produce full sentences. What is significant is that as long and hard as it has been, she has shown amazing progress and remarkable improvement over a sustained period of time. Persistence, hard work, support, and determination characterize the story of Gabby Giffords.

Language improvement in aphasia can and does occur over time. It may not happen in a day, a week, a month, or even a year. And it typically does not result in a full recovery of language at a level prior to the injury. This is an important message for all of those who have aphasia and for their families. There is no guarantee about what or when the end-state of the recovery process will be. Change may happen imperceptibly and gradually, or not at all. Still, the message is to *never give up and give it time* – keep working, talking, and communicating.

### 9.1.3  Words and Music

Music holds a special place in aphasia. It can bring out language when it seems to be elusive. It is not uncommon for a patient who cannot speak or who struggles to get words out to be able to produce the lyrics of a song while singing the melody. Both the words and the music are of a piece. In addition to providing a venue for speaking, it is a concrete example to the patient and family that language is there. It has not disappeared.

One patient whom I have known for many years has a severe Broca's aphasia. Although his speech has improved over time, he has disfluencies as he searches for words. His speech comes out with difficulty and the sounds are often distorted. In sharp contrast, he has a fine singing voice, and when he sings, the words are produced clearly and fluently without any sign of articulatory difficulties. I remember him singing a well-known song during a demonstration for a class in neuroscience. The audience, made up of students and faculty, was astounded not only by the richness of his singing voice, but by the change in the quality of his speech when he sang compared to when he just talked. His enunciation was perfect; there were no disfluencies and he articulated the words easily and clearly.

This connection between sounds and music has been harnessed in a speech therapy program called Melodic Intonation Therapy (MIT). Created in 1973, it has been one of the most successful therapy programs in helping those with severe nonfluent aphasia to say individual words and to begin to string words together. Here, the elements of music, melody and rhythm, are yoked to words. Typically, the program starts with single words and gradually increases productions to longer strings. For example, the patient might be asked to intone a short sentence like 'my name is ___'. It is not necessary that a particular tune or melody be used for any particular production nor is it required that all of the words be produced. Rather, the patient's productions are to be articulated in a sing-song manner.

This procedure has shown a modest increase in the ability of those with nonfluent aphasia to begin to produce single words and to start to put words together. Why this program is so effective is the subject of much discussion in the aphasia literature. Some suggest that MIT engages right hemisphere mechanisms that are recruited in the processing of music. Others suggest that the rhythm and melody of music enhance access to language in the left hemisphere. Suffice to say, at the minimum, MIT therapy provides a promising first step in starting the patient towards being able to produce language.

### 9.1.4   Insight

Many people who meet individuals with aphasia wrongly assume that because they struggle to use language, they are cognitively impaired, fail to understand their situation, have minimal awareness of their disorder, and are unable to think, feel, or have normal emotions. It is not uncommon for family and friends to underestimate the capacity of the patient and accommodate their behavior by speaking loudly, as though the person with aphasia cannot hear, using simplified, language, even talking 'baby talk'. In truth, it is hard to know what is going on 'inside the head' of someone with aphasia at the acute, early stages of the disorder. However, as the patient improves and is able to interact more easily, examples abound showing that there is an awareness and often insight into the disability.

Years ago, I was working with a young man with aphasia. He had a classic Broca's aphasia – his speech output was slow and agrammatic and his comprehension was good. He was describing his difficulty in speaking and he showed remarkable awareness of his impairment. He said (and I paraphrase) 'words, … little words, I know, … I don't know'. He perfectly indicated that he was having trouble with the 'little words' like 'the', 'a', 'and', 'be', just the words that are typically omitted when he was speaking and when someone with Broca's aphasia is agrammatic. Even those with Wernicke's aphasia may show uncommon 'understanding', despite using jargon in speech output and having a severe auditory comprehension problem. One patient referred to his speech as 'rough speak', an astute observation of the hallmark of his speech output disorder.

We generally consider an aphasic 'cured' when they no longer show impairments on standardized aphasia tests like the Boston Diagnostic Aphasia Exam (Goodglass and Kaplan 1983) or the Western Aphasia Battery (Kertesz 1982). Despite what might appear as 'normal' language to the outside world, it is not uncommon for the aphasic to be aware of lingering difficulties and challenges while trying to navigate the world as they did prior to their disorder.

C. Scott Moss incisively describes this in his book *Recovery with Aphasia: The Aftermath of My Stroke* (1972). A psychologist, who taught and did research, he found giving spontaneous lectures to his class, responding to unanticipated questions, talking on the fly with others, responding to changes in topics quickly, remained difficult for him even after his language was clinically assessed as completely recovered.

Language communication is complex. It is more than simply producing well-articulated words and sentences. Under normal circumstances, daily communication is varied. It requires the mental and linguistic agility to change topics, to say things in new ways, and to not rely on stock, stereotypical phrases. For those who appear to be recovered, there may still be an awareness that all is not right. Such individuals may be frustrated by this circumstance, and may have to make adjustments to their life circumstances. Moss did just that. Because he found teaching, advising students, and doing research to be too difficult and challenging, he ultimately left his position to work in an environment that allowed for greater 'control' of his social context and ultimately how he used language. He successfully made that adjustment.

Moss' circumstances should not be taken as a depressing or a bad outcome. Rather, he and his family had realistic goals and made the necessary adjustments to allow him to continue to lead a rich and productive life just as Patricia Neal did and as Gabby Giffords is now doing. Their stories are a testament to the strength of the human spirit. Together, they show that although the road to recovery in aphasia may be long and arduous and may not be complete, each of them met the challenge in their own way. The extent of recovery is an unknown. That is a sobering fact. However, recognizing that there may be limits is not a recipe for giving up. Always the goal should be to provide all possible conditions that will facilitate language and communication including continued and consistent support from family and friends, advocacy for the best possible treatments and therapy, patience, optimism, hope, and persistence.

## 9.2  A Working Agenda for the Future

We end this book on a message of hope. There are lessons learned not just from the science of aphasia but also from the human stories of aphasia. And there is much more to learn. So the story has not ended. While some answers have been provided, there are many more questions remaining. This may be frustrating for you. But this is the essence of scientific investigations. Even when they provide explanations, they typically raise more questions. So as this book comes to a close, the question is where do we go from here? How do we use what we have learned and move forward in our understanding of aphasia and in providing the best means of helping those with aphasia to compensate for and overcome their language impairment? The goal then of the remaining portion of this book is to use what we have learned to frame an agenda for the future.

Perhaps one of the greatest challenges we face is connecting the results from basic science to the patient. Indeed, as in many fields of study, there is often a disconnect between these two areas. And the problem goes in both directions. The world of aphasia is no different. The goals, methods, and tools used in basic science research and its applications to real world problems, while related (both are trying to answer questions that speak to the neural basis of language) are different. Typically, the goal of the scientist is to provide an 'explanation' for a particular phenomenon by focusing on a question, developing a hypothesis, testing it experimentally, and drawing conclusions based on the resulting data. The scientific results may have no clear-cut application to a 'real world' problem. In contrast, the job of the therapist is to deal with the real world problem; to understand the difficulties of the patient and to develop a therapy program that will lead to recovery. But those goals, at least on the surface, are different and hence the two fields often do not contact each other. The result? In the worst case, there is a failure to utilize what we know about the science of aphasia and the science of the brain in developing and implementing the best possible therapy programs.

Here it may sound like the 'fault' in developing programs that are based on state-of-the-art knowledge of aphasia lies with those who do therapy. This is not the case. To solve a problem is a two way street – one that requires each area to be aware of and learn and understand at the minimum the basics of the other area. And each area shares some of the blame for not being cognizant of the other area.

Let's first consider those who focus on the science of aphasia. Over the last 10 years, basic research on the cognitive neuroscience of language has gradually turned away from a focus on aphasia to explore basic science questions using the new methods described in Chap. 8 such as fMRI, PET, and TMS. These methods provide better localization than lesion analyses and allow for close examination of the neural systems underlying language in the uninjured brain. For these researchers, the focus is on the processes and mechanisms supporting language and the neural systems underlying them. How this relates to individuals who sustain brain injury or the implications such injury has for their recovery is not part of the equation. Indeed, many of these researchers know little about aphasia and its clinical (behavioral) manifestations.

For those who work with brain-injured individuals with aphasia, there is often minimal connection to and familiarity with findings in the science of aphasia as well as in investigations of the processes and mechanisms involved in speaking and understanding absent brain injury. This means that basic research tends not to influence therapy programs. In truth, every science uses specialized vocabulary or jargon, relies on theoretical assumptions and models, and applies detailed experimental methods. The result is that it is a hard slog at best for those not working directly in the field to follow the literature and to consider how the findings might inform therapy and ultimately be applied to individuals with aphasia.

All of this sounds very negative and depressing. But it need not be. There is a path forward, one that requires the integration of basic and applied science by conducting research that aims specifically at translating research findings directly to the patient. Although the literature is not extensive, there are a number of researchers focusing on aphasia who are bridging this gap. There is a full agenda which can be mapped out. I will briefly identify three possible areas ripe for intensive study. The first focuses on an approach to speech therapy that is grounded in what we know about how language is learned. Here, we will consider how basing therapy on principles of learning may enhance speech and word processing. The last two are more speculative as they focus on whether exciting or inhibiting neural connections will serve to enhance language recovery.

## 9.2.1    From Bench to Bedside

### 9.2.1.1 Therapy Programs Based on Basic Science Findings

Speech therapy played a large role in each of the vignettes we have just talked about and it is a critical part of the recovery process in aphasia. It is worthwhile to consider how such therapy programs are developed. Rather than using the same therapy regimen for each patient, the type of therapy is typically targeted and tailored to the needs of the patient. It makes sense to use a different approach for those patients who understand language but struggle to speak than for those patients who do not understand language but who speak fluently and copiously but with empty content. But what is the right approach and how does one decide?

Common sense can be used, in part, as a means of focusing on different aspects of language that need to be worked on. But surely, there must be more to it. And there is. Consider what you have learned about aphasia and the patterns and types of deficits resulting from brain injury. Here, we applied the scientific method looking at findings that had been obtained by systematically observing and collecting data, and developing and testing hypotheses with the goal of seeking explanations. We know that there are similarities in the way language breaks down across patients and, as well, there are differences that emerge depending on the areas of brain injury. We know that some aspects of language are more vulnerable than others, and the fundamentals of language, its architecture, appear to be spared in aphasia. Taken together, these findings provide the path to go from discovery in the lab (the *bench*) to therapy for the patient (the *bedside*).

Let's go back to MIT therapy for a moment. Here, the therapists take advantage of something that is preserved in Broca's aphasia, the ability to produce words while singing, to work on the heart of the patient's impairment, an inability to produce speech in conversation or in response to a question. Music then is used as a proxy for communicating with speech. In this case, MIT is serving as a wedge into the language impairment of the patient. It uses what is 'right' as a scaffold to 'bootstrap' and help access those aspects of language that are impaired.

Music is not the only entry point in developing a therapy program aimed at improving speech production. As we discussed in Chap. 1, aphasia syndromes reflect a constellation of impaired as well as spared abilities. There is also a range of severity of impairment among the different language abilities. For example, a patient may do better repeating what the examiner says than spontaneously speaking with someone. If that is the case, a beginning strategy might be to have the patient repeat a phrase such as 'I am fine' and then segue to using that phrase in response to a question such as 'how are you?' Typically, such a strategy will not work overnight and will take work and practice. But it is a start.

Nonetheless, as important as these common sense strategies are, there is much we have learned from basic research on language processing in the uninjured brain that can serve as a framework for guiding therapy programs more generally. Indeed, in some cases what we have learned would seem to go against common sense.

Let's be more specific. We know that those with aphasia have difficulty producing the sounds of language, often substituting one sound for another. And in the case of those with Broca's aphasia the quality of the sound productions are often distorted. We also know that those with aphasia often have difficulty naming objects, often substituting the name of an object that shares meaning. How would you approach therapy in these instances?

Your intuition might be to start simple and increase complexity over time. In speech production, one might ask the patient to produce or repeat the same syllable such as 'pa' multiple times, and then move on to producing that sound in a different vowel context such as 'pi' or producing a similar consonant sound such as 'ba'. And in naming, the focus might be on presenting the patient a picture of an object to be named and then move on to the next picture representing a different object. The process would be repeated multiple times. In both cases, one exemplar of a sound is presented and one exemplar of a picture is presented. However, basic science research has shown that simple is not always the answer. Rather, variability is the key to learning the categories of sounds and objects. Why is this the case?

In both speech production and perception, we know that different speakers and different vowel contexts result in variability in both the production and perception of sounds. There is a range of variants of the same sound that belong to the same sound category. This range of variants facilitates the learning of new speech sounds when learning a second language because it gives both our perceptual and production systems a framework in which to place the sound.

Think about it. Presenting only one example of a speech sound is just that, one example. It does not provide the listener or the speaker with the range of possibilities of a given sound and the boundary between different sounds, making it impossible to know when a particular sound such as 'p' is a variant of the same sound in 'p$^1$ear' vs. 'p$^2$ear' or when a sound belongs to a different category such as 'pear' vs. 'bear'. So even though it may seem counterintuitive, variability is a help, not a hindrance, when working on improving the perception and production of sounds in aphasia.

It is the same for objects. We recognize the same object even if it is seen from different viewpoints. A cup is still a cup whether seen upside down or right side up. And we recognize different objects as belonging to the same category despite variability in their dimensions or in some of their features. There are big dogs and little dogs. Size in this case does not matter. But a dog becomes a cat when there is a change of features of the ears and face. Here again, a single exemplar does not provide enough information about what features are relevant and what features are not for membership in a particular category. Variability, not simplicity, is the key for naming a category and for presumably 'reactivating' its features in aphasia.

### 9.2.1.2 Mapping Neural Changes in the Most Effective Therapy Programs

Some therapy programs may have greater success than others in the recovery of language in aphasia. Why? The content of the therapy program certainly is a critical factor. However, there are neural consequences of therapy as well. We know that there are neural changes as language improves. What we don't know is whether there are different effects of these programs on the neural areas recruited or changes in the connections between areas that are predictive of good behavioral outcomes. We also don't know whether the improved language abilities rely on language areas, non-language areas, or even new connections to areas not typically engaged in speaking and understanding. Identifying those areas that enhance language recovery may serve as 'targets' for neural rehabilitation through a range of stimulation techniques. Such techniques could include presenting to the patient behavioral tasks that are known to activate those particular areas or by brain stimulation techniques such as anodal transcranial direct current stimulation (tDCS) which excite selected neural areas.

Consider, for example, melodic intonation therapy which we described earlier in this chapter. MIT therapy enhances speech output. What we don't know is whether MIT recruits language and/or non-language areas, or whether it engages the right hemisphere or depends on left hemisphere resources. Knowing the neural areas recruited during MIT as well as their function more broadly may provide 'targets' for neural stimulation. And doing so may serve to excite the speech production network to ultimately enhance speech output in aphasia.

## 9.3   Let's Finally Figure Out the Right Hemisphere

We have spent a lot of time talking about the right hemisphere in Chap. 8 and earlier in this chapter. Why so much time and space? Knowing the role of the right hemisphere is critical for understanding the potential role it may play in recovery of language in aphasia. What is the relation between improvement in therapy and the engagement of the right hemisphere?

As we described earlier, the engagement of the right hemisphere appears to vary over time in those recovering from aphasia. There is greater activation in the early stages of recovery and gradually less activation as language improves. Knowing the time course of this involvement and whether the right hemisphere is helping or hindering recovery is a crucial question.

If we assume that the right hemisphere is actively contributing to language post-stroke, then stimulating it using techniques such as TMS might enhance its ability to contribute to language improvement. Such stimulation may need to occur throughout the time course of recovery in order to keep the right hemisphere actively engaged. In contrast, if the right hemisphere is impeding progress, then inhibiting its activation during recovery may have the counterintuitive but positive effect of improving language performance.

It is possible that the right hemisphere's role in language may vary depending on the components of language involved. For example, it may enhance recovery in accessing meaning, and it may negatively affect recovery in speech production. In such cases, the type of stimulation as well as the neural areas stimulated will need to vary depending on what aspect of language is being targeted.

## 9.4   The Circle Closes

We have now come full circle and completed our journey into how mind meets language in the brain through the lens of aphasia. We have focused largely on the science of aphasia, always keeping in mind that the science is never divorced from the human realities of this disorder. Studied together, they allow for a deeper understanding of what happens when brain injury affects the ability to speak and understand. Much remains to be done. My hope is that you have learned a lot in reading this book, and that it has sparked in you an abiding interest in aphasia and in the brain and language. I also hope that some of you will take up the mantle to answer the many questions raised and to find a solution to what happens *when words betray us*.

## References

Farrell, B. (1969). Pat and Roald. New York: Random House.

Giffords, G. and Kelly, M. (2012). Gabby: A Story of Courage, Love, and Resilience. New York: Scribner.

Goodglass, H. and Kaplan, E. (1983). The assessment of aphasia and related disorders. 2nd Edition. Philadeplphia: Lea and Febiger.

Kertesz, A. (1982). The Western Aphasia Battery. New York: Grune & Stratton, Inc.

Moss, C. Scott. (1972). Recovery with Aphasia: The Aftermath of My Stroke. Urbana: University of Illinois Press).

## Readings of Interest

Albert, M. L., Sparks, R. W., and Helm, N. A. (1973). Melodic intonation therapy for aphasia. Archives of Neurology, 29(2), 130–131.

Kiran, S. (2009). Complexity in the treatment of naming deficits. American Journal of Speech-Language Pathology. American Journal of Speech and Lang Pathology, 16(1), 18–29. doi: https://doi.org/10.1044/1058-0360(2007/004)

Kiran, S., and Thompson, C. K. (2003). The role of semantic complexity in treatment of naming deficits: training semantic categories in fluent aphasia by controlling exemplar typicality. Journal of Speech, Language, and Hearing Research, 46(4), 773–787. doi:https://doi.org/10.1044/1092-4388(2003/061)

Sacks, O. (2010). Recalled to Life. In The Mind's Eye. New York: A.A. Knopf, pp. 32–52.

Sparks, R., Helm, N., and Albert, M. (1974). Aphasia rehabilitation resulting from melodic intonation therapy. Cortex, 10(4), 303–316.

Thompson, C. K. (2019). Neurocognitive recovery of sentence processing in aphasia. Journal of Speech, Language, and Hearing Research, 62(11), 3947–3972. doi:https://doi.org/10.1044/2019_JSLHR-L-RSNP-19-0219

# Index

© Springer Nature Switzerland AG 2022
S. E. Blumstein, *When Words Betray Us*,
https://doi.org/10.1007/978-3-030-95848-0

Printed in the United States
by Baker & Taylor Publisher Services